DATA SCIENCE
FOR UNDERGRADUATES
OPPORTUNITIES AND OPTIONS

Committee on Envisioning the Data Science Discipline:
The Undergraduate Perspective

Computer Science and Telecommunications Board
Board on Mathematical Sciences and Analytics
Committee on Applied and Theoretical Statistics
Division on Engineering and Physical Sciences

Board on Science Education
Division of Behavioral and Social Sciences and Education

A Consensus Study Report of

The National Academies of
SCIENCES • ENGINEERING • MEDICINE

THE NATIONAL ACADEMIES PRESS
Washington, DC
www.nap.edu

THE NATIONAL ACADEMIES PRESS 500 Fifth Street, NW Washington, DC 20001

This activity was supported by Award No. 1626983 from the National Science Foundation (Directorate for Computer and Information Science and Engineering; Directorate for Education and Human Resources; Directorate for Mathematical and Physical Sciences/Division of Mathematical Sciences; and Directorate for Social, Behavioral and Economic Sciences). Any opinions, findings, conclusions, or recommendations expressed in this publication do not necessarily reflect the views of any organization or agency that provided support for the project.

International Standard Book Number-13: 978-0-309-47559-4
International Standard Book Number-10: 0-309-47559-7
Digital Object Identifier: https://doi.org/10.17226/25104

Additional copies of this publication are available for sale from the National Academies Press, 500 Fifth Street, NW, Keck 360, Washington, DC 20001; (800) 624-6242 or (202) 334-3313; http:// www.nap.edu.

Copyright 2018 by the National Academy of Sciences. All rights reserved.

Printed in the United States of America

Suggested citation: National Academies of Sciences, Engineering, and Medicine. 2018. *Data Science for Undergraduates: Opportunities and Options*. Washington, DC: The National Academies Press. https://doi.org/10.17226/25104.

The National Academies of
SCIENCES · ENGINEERING · MEDICINE

The **National Academy of Sciences** was established in 1863 by an Act of Congress, signed by President Lincoln, as a private, nongovernmental institution to advise the nation on issues related to science and technology. Members are elected by their peers for outstanding contributions to research. Dr. Marcia McNutt is president.

The **National Academy of Engineering** was established in 1964 under the charter of the National Academy of Sciences to bring the practices of engineering to advising the nation. Members are elected by their peers for extraordinary contributions to engineering. Dr. C. D. Mote, Jr., is president.

The **National Academy of Medicine** (formerly the Institute of Medicine) was established in 1970 under the charter of the National Academy of Sciences to advise the nation on medical and health issues. Members are elected by their peers for distinguished contributions to medicine and health. Dr. Victor J. Dzau is president.

The three Academies work together as the **National Academies of Sciences, Engineering, and Medicine** to provide independent, objective analysis and advice to the nation and conduct other activities to solve complex problems and inform public policy decisions. The National Academies also encourage education and research, recognize outstanding contributions to knowledge, and increase public understanding in matters of science, engineering, and medicine.

Learn more about the National Academies of Sciences, Engineering, and Medicine at www.nationalacademies.org.

The National Academies of
SCIENCES • ENGINEERING • MEDICINE

Consensus Study Reports published by the National Academies of Sciences, Engineering, and Medicine document the evidence-based consensus on the study's statement of task by an authoring committee of experts. Reports typically include findings, conclusions, and recommendations based on information gathered by the committee and the committee's deliberations. Each report has been subjected to a rigorous and independent peer-review process and it represents the position of the National Academies on the statement of task.

Proceedings published by the National Academies of Sciences, Engineering, and Medicine chronicle the presentations and discussions at a workshop, symposium, or other event convened by the National Academies. The statements and opinions contained in proceedings are those of the participants and are not endorsed by other participants, the planning committee, or the National Academies.

For information about other products and activities of the National Academies, please visit www.nationalacademies.org/about/whatwedo.

COMMITTEE ON ENVISIONING THE DATA SCIENCE DISCIPLINE: THE UNDERGRADUATE PERSPECTIVE

LAURA HAAS, NAE,[1] University of Massachusetts Amherst, *Co-Chair*
ALFRED O. HERO III, University of Michigan, *Co-Chair*
ANI ADHIKARI, University of California, Berkeley
DAVID CULLER, NAE, University of California, Berkeley
DAVID DONOHO, NAS,[2] Stanford University
E. THOMAS EWING, Virginia Polytechnic Institute and State University
LOUIS J. GROSS, University of Tennessee, Knoxville
NICHOLAS J. HORTON, Amherst College
JULIA LANE, New York University
ANDREW McCALLUM, University of Massachusetts Amherst
RICHARD McCULLOUGH, Harvard University
REBECCA NUGENT, Carnegie Mellon University
LEE RAINIE, Pew Research Center
ROB RUTENBAR, University of Pittsburgh
KRISTIN TOLLE, Microsoft Research
TALITHIA WILLIAMS, Harvey Mudd College
ANDREW ZIEFFLER, University of Minnesota, Minneapolis

Staff

MICHELLE K. SCHWALBE, Director, Board on Mathematical Sciences and Analytics (BMSA), *Study Director*
JON EISENBERG, Director, Computer Science and Telecommunications Board (CSTB)
BEN WENDER, Director, Committee on Applied and Theoretical Statistics
AMY STEPHENS, Program Officer, Board on Science Education
LINDA CASOLA, BMSA, Associate Program Officer and Editor
RENEE HAWKINS, CSTB, Financial Manager
JANKI PATEL, CSTB, Senior Program Assistant

---

[1] Member, National Academy of Engineering.
[2] Member, National Academy of Sciences.

COMPUTER SCIENCE AND TELECOMMUNICATIONS BOARD

FARNAM JAHANIAN, Carnegie Mellon University, *Chair*
LUIZ ANDRÉ BARROSO, Google, Inc.
STEVEN M. BELLOVIN, NAE,[1] Columbia University
ROBERT F. BRAMMER, Brammer Technology, LLC
DAVID CULLER, NAE, University of California, Berkeley
EDWARD FRANK, Cloud Parity, Inc.
LAURA HAAS, NAE, University of Massachusetts Amherst
MARK HOROWITZ, NAE, Stanford University
ERIC HORVITZ, NAE, Microsoft Corporation
VIJAY KUMAR, NAE, University of Pennsylvania
BETH MYNATT, Georgia Institute of Technology
CRAIG PARTRIDGE, Raytheon BBN Technologies
DANIELA RUS, NAE, Massachusetts Institute of Technology
FRED B. SCHNEIDER, NAE, Cornell University
MARGO SELTZER, Harvard University
MOSHE VARDI, NAS[2]/NAE, Rice University

Staff

JON EISENBERG, Director
LYNETTE I. MILLETT, Associate Director
EMILY GRUMBLING, Program Officer
KATIRIA ORTIZ, Associate Program Officer
RENEE HAWKINS, Financial and Administrative Manager
JANKI PATEL, Senior Program Assistant
SHENAE BRADLEY, Administrative Assistant

---

[1] Member, National Academy of Engineering.
[2] Member, National Academy of Sciences.

BOARD ON MATHEMATICAL SCIENCES AND ANALYTICS

STEPHEN M. ROBINSON, NAE,[1] University of Wisconsin, Madison, *Chair*
JOHN R. BIRGE, NAE, University of Chicago
W. PETER CHERRY, Independent Consultant
DAVID CHU, Institute for Defense Analyses
RONALD R. COIFMAN, NAS,[2] Yale University
JAMES CURRY, University of Colorado Boulder
CHRISTINE FOX, Johns Hopkins Applied Physics Laboratory
MARK L. GREEN, University of California, Los Angeles
PATRICIA A. JACOBS, Naval Postgraduate School
JOSEPH A. LANGSAM, Morgan Stanley (Retired)
SIMON A. LEVIN, NAS, Princeton University
ANDREW W. LO, Massachusetts Institute of Technology
DAVID MAIER, Portland State University
LOIS CURFMAN McINNES, Argonne National Laboratory
FRED S. ROBERTS, Rutgers, The State University of New Jersey
ELIZABETH A. THOMPSON, NAS, University of Washington
CLAIRE TOMLIN, University of California, Berkeley
LANCE WALLER, Emory University
KAREN WILLCOX, Massachusetts Institute of Technology
DAVID YAO, NAE, Columbia University

Staff

MICHELLE K. SCHWALBE, Director
BEN WENDER, Program Officer
LINDA CASOLA, Associate Program Officer and Editor
BETH DOLAN, Financial Manager
RODNEY N. HOWARD, Administrative Assistant

---

[1] Member, National Academy of Engineering.
[2] Member, National Academy of Sciences.

COMMITTEE ON APPLIED AND THEORETICAL STATISTICS

ALFRED O. HERO III, University of Michigan, *Chair*
ALICIA CARRIQUIRY, NAM,[1] Iowa State University
MICHAEL J. DANIELS, University of Florida
KATHERINE BENNETT ENSOR, Rice University
AMY HERRING, Duke University
NICHOLAS J. HORTON, Amherst College
DAVID MADIGAN, Columbia University
JOSÉ M.F. MOURA, NAE,[2] Carnegie Mellon University
NANCY REID, NAS,[3] University of Toronto
CYNTHIA RUDIN, Duke University
AARTI SINGH, Carnegie Mellon University

Staff

BEN WENDER, Director
LINDA CASOLA, Associate Program Officer and Editor
BETH DOLAN, Financial Manager
RODNEY N. HOWARD, Administrative Assistant

---

[1] Member, National Academy of Medicine.
[2] Member, National Academy of Engineering.
[3] Member, National Academy of Sciences.

BOARD ON SCIENCE EDUCATION

ADAM GAMORAN, William T. Grant Foundation, *Chair*
SUNITA V. COOKE, MiraCosta College
MELANIE COOPER, Michigan State University
RODOLFO DIRZO, NAS,[1] Stanford University
RUSH D. HOLT, American Association for the Advancement of Science
MATTHEW KREHBIEL, Achieve, Inc.
MICHAEL LACH, University of Chicago
LYNN LIBEN, Pennsylvania State University
CATHRYN (CATHY) MANDUCA, Carleton College
JOHN MATHER, NAS, NASA Goddard Space Flight Center
TONYA M. MATTHEWS, Michigan Science Center
BRIAN REISER, Northwestern University
MARSHALL (MIKE) SMITH, Carnegie Foundation for the Advancement of Teaching
ROBERTA TANNER, Thompson School District (Retired)
SUZANNE WILSON, Michigan State University

Staff

HEIDI SCHWEINGRUBER, Director
KERRY BRENNER, Senior Program Officer
MARGARET HILTON, Senior Program Officer
KENNE DIBNER, Program Officer
AMY STEPHENS, Program Officer
MATTHEW LAMMERS, Program Coordinator
LETICIA GARCILAZO GREEN, Senior Program Assistant
MARGARET KELLY, Senior Program Assistant
COREETHA ENTZMINGER, Program Assistant

---

[1] Member, National Academy of Sciences.

Preface

The National Academies of Sciences, Engineering, and Medicine established the Committee on Envisioning the Data Science Discipline: The Undergraduate Perspective to set forth a vision for the emerging discipline of data science at the undergraduate level (see Box P.1 for the committee's statement of task).

This study was sponsored by the National Science Foundation. The Committee on Envisioning the Data Science Discipline: The Undergraduate Perspective (see Appendix A for biographical sketches of the committee members) conducted a number of information-gathering activities and engaged a broad community in its conversations to address the statement of task shown in Box P.1 (see Appendix B for a list of the presentations given during these meetings and Appendix C for a list of those who contributed). In December 2016, the committee met to discuss possible future directions based on progress with current data science programs; societal implications of the evolving field of data science; approaches to expand diversity and inclusion in data science among students, staff, and topic areas; and perspectives on envisioning the future of data science for undergraduates. In April 2017, the committee organized a webinar to collect further input from the public on topics of importance for this study.

In May 2017, the committee convened a workshop in which participants discussed educational models to build relevant foundational, translational, and professional skills for data scientists in various roles; the use of high-impact educational practices in the delivery of data science education; and strategies for broad participation in data science educa-

> **BOX P.1**
> **Statement of Task**
>
> A National Academies of Sciences, Engineering, and Medicine study will set forth a vision for the emerging discipline of data science at the undergraduate level. It will emphasize core underlying principles, intellectual content, and pedagogical issues specific to data science, including core concepts that distinguish it from neighboring disciplines. It will not consider the practicalities of creating materials, courses, or programs. It will develop this vision considering applications of and careers in data science. It will focus on the undergraduate level, addressing related issues at the middle and high school level as well as community colleges as appropriate, and will draw on experiences in creating master's-level programs. It will also consider opportunities created by the emergence of a new science, technology, engineering, and mathematics (STEM) field to engage underrepresented student populations and consider ways to reduce the "leakage" seen in existing STEM pathways. Information gathering will center around two workshops, the first likely focused on principles and intellectual content, and the second likely focused on pedagogy and implications for middle and high schools and community colleges. To get material on the record quickly and spark community feedback, a rapporteur-authored workshop summary report will be issued following the first workshop. A final report will be issued following both workshops and committee deliberations setting forth a vision for undergraduate education in data science.

tion that rely on formal modes of evaluation and assessment. Participants focused on the ways in which students, institutions, and programs could change in the coming decade, as well as how these changes will affect future plans for data science education.

The committee also held nine webinars throughout fall 2017 as another means to engage the public in conversations about various aspects of data science education, which addressed the following topics:

1. Building data acumen;
2. Incorporating real-world applications;
3. Training faculty and developing curriculum;
4. Enhancing communication and teamwork skills;
5. Fostering interdepartmental collaboration and institutional organization;
6. Considering ethics;
7. Assessing and evaluating data science programs;
8. Emphasizing diversity, inclusion, and increased participation; and
9. Exploring 2-year colleges and institutional partnerships.

Although these nine webinars focused specifically on applications to data science programs, many of the discussions highlighted best practices relevant for all types of academic programming. The committee met for a final session in December 2017 to prepare for the writing of this report. During this session, the committee synthesized discussions from the webinar series and results from activities under way in the data science community. This final report, which was preceded by a September 2017 interim report, explores key questions about the future of the field of data science.

Acknowledgments

This Consensus Study Report has been reviewed in draft form by individuals chosen for their diverse perspectives and technical expertise. The purpose of this independent review is to provide candid and critical comments that will assist the National Academies of Sciences, Engineering, and Medicine in making each published report as sound as possible and to ensure that it meets the institutional standards for quality, objectivity, evidence, and responsiveness to the study charge. The review comments and draft manuscript remain confidential to protect the integrity of the deliberative process.

We thank the following individuals for their review of this report:

Richard (Dick) De Veaux, Williams College,
Natalie M. Evans Harris, BrightHive,
Charles Isbell, Jr., Georgia Institute of Technology,
Iain Johnstone, NAS,[1] Stanford University,
Brian Kotz, Montgomery College,
Peter Norvig, Google, Inc.,
Renata Rawlings-Goss, South Big Data Regional Innovation Hub and Georgia Institute of Technology,
Ali Sayed, NAE,[2] University of California, Los Angeles,
Margo Seltzer, Harvard University, and
Sharon Wood, NAE, University of Texas, Austin.

---

[1] Member, National Academy of Sciences.
[2] Member, National Academy of Engineering.

Although the reviewers listed above provided many constructive comments and suggestions, they were not asked to endorse the conclusions or recommendations presented in the report, nor did they see the final draft of the report before its release. The review of this report was overseen by Alicia L. Carriquiry, NAM,[3] Iowa State University. She was responsible for making certain that an independent examination of this report was carried out in accordance with the standards of the National Academies and that all review comments were carefully considered. Responsibility for the final content of this report rests entirely with the authoring committee and the National Academies.

The committee would like to thank Andy Burnett from Knowinnovation for facilitating the committee's May 2017 workshop as well as the following staff members from the National Science Foundation for their input, assistance, and support of this study: Stephanie August, Chaitan Baru, Eva Campo, Vandana Janeja, Nandini Kannan, Sara Kiesler, Gabriel Perez-Giz, Earnestine Psalmonds-Easter, and Elena Zheleva. The committee would also like to thank the many individuals who provided input to this study; the full list of these individuals is included in Appendix C.

---

[3] Member, National Academy of Medicine.

Contents

SUMMARY		1
1	INTRODUCTION A Look to the Future, 9 Report Overview, 10 References, 10	6
2	KNOWLEDGE FOR DATA SCIENTISTS Data Scientists of Today and Tomorrow, 17 Data Acumen, 21 A Code of Ethics for Data Science, 31 References, 33	12
3	DATA SCIENCE EDUCATION Undergraduate Modalities, 35 Middle and High School Education, 56 References, 58	35
4	STARTING A DATA SCIENCE PROGRAM Ensuring Broad Participation, 61 Academic Infrastructure, 65 Curriculum, 68 Faculty Resources, 69 Assessment, 70 References, 71	60

5	**EVOLUTION AND EVALUATION** Evolution, 73 Evaluation, 78 Roles for Professional Societies, 83 References, 85	72
6	**CONCLUSIONS**	87

APPENDIXES

A	Biographies of the Committee	95
B	Meetings and Presentations	105
C	Contributing Individuals	111
D	Data Science Oath	117

Summary

Data science is emerging as a field that is revolutionizing science and industries alike. Work across nearly all domains is becoming more data driven, affecting both the jobs that are available and the skills that are required. As more data and ways of analyzing them become available, more aspects of the economy, society, and daily life will become dependent on data. As a result, the National Academies of Sciences, Engineering, and Medicine were asked to set forth a vision for the emerging discipline of data science at the undergraduate level. To that end, the committee considered core underlying principles, intellectual content, and pedagogical issues specific to data science, including the essential concepts that distinguish it from neighboring disciplines. All of this was anchored in exploration related to applications of and careers in data science.

Today, the term "data scientist" typically describes a knowledge worker who is principally occupied with analyzing complex and massive data resources. However, data science spans a broader array of activities that involve applying principles for data collection, storage, integration, analysis, inference, communication, and ethics. In future decades, all undergraduates will benefit from a fundamental awareness of and competence in data science.

Recommendation 2.3: To prepare their graduates for this new data-driven era, academic institutions should encourage the development of a basic understanding of data science in all undergraduates.

The continued transformation of work requires both a larger population with a basic understanding of data science and a substantial cadre of talented graduates with highly developed data science skills and knowledge, acquired through substantial coursework and practice.

> **Recommendation 2.1: Academic institutions should embrace data science as a vital new field that requires specifically tailored instruction delivered through majors and minors in data science as well as the development of a cadre of faculty equipped to teach in this new field.**

The new majors and minors will initially combine ingredients from existing courses, in areas such as computer science, statistics, business analytics, information technology, optimization, applied mathematics, and numerical computing. Over time, as features of the new data-driven era take shape, academic programs will be compelled to develop new skill clusters, and a body of distinctive courses and instructional materials will emerge.

> **Recommendation 4.1: As data science programs develop, they should focus on attracting students with varied backgrounds and degrees of preparation and preparing them for success in a variety of careers.**

Graduates of these programs will work in virtually every job sector and will serve in a number of roles, including operating the systems on which analyses are run, preparing data for analysis, defining and coordinating the analysis, visualizing information, and supporting data-driven decision making to uncover the stories buried in the data. Others who use data science skills will be journalists, administrators, artists, lawyers, teachers, and other workers who need some ability to understand and use data. This need to prepare diverse students for various careers further increases the educational challenge.

A wide variety of instructional programs will be needed to prepare students for the data-enriched world of the coming years.

> **Recommendation 2.2: Academic institutions should provide and evolve a range of educational pathways to prepare students for an array of data science roles in the workplace.**

These include introductory courses, full degrees at both associate and bachelor levels, and a range of minors and certificates. The forms of these programs and their scope will vary depending on the culture of a given institution and the aims of its students.

Regardless of the type of program, certain elements need to be covered, though perhaps to varying degrees and with varying emphases. A key goal is to give all students the ability to make good judgments, use tools responsibly and effectively, and ultimately make good decisions using data. The committee defines this collection of abilities as "data acumen." To that end, students will need exposure to material from multiple disciplines—notably, mathematical, statistical, and computational foundations—and they will need training in data acquisition, modeling, management and curation, data visualization, workflow and reproducibility, communication and teamwork, domain-specific considerations, and ethical problem solving.

The committee underscores the centrality of studying the many ethical considerations that arise as workers engage in data science. These considerations include deciding what data to collect, obtaining permissions to use data, crediting the sources of data properly, validating the data's accuracy, taking steps to minimize bias, safeguarding the privacy of individuals referenced in the data, and using the data correctly and without alteration. It is important that students learn to recognize ethical issues and to apply a high ethical standard.[1]

Recommendation 2.4: Ethics is a topic that, given the nature of data science, students should learn and practice throughout their education. Academic institutions should ensure that ethics is woven into the data science curriculum from the beginning and throughout.

Recommendation 2.5: The data science community should adopt a code of ethics; such a code should be affirmed by members of professional societies, included in professional development programs and curricula, and conveyed through educational programs. The code should be reevaluated often in light of new developments.

Academic institutions are stepping up to these educational challenges with a variety of programs and educational pathways. Several 4-year undergraduate institutions offer data science majors and/or minors—serving not only those students pursuing data science as a career but also those students who want to acquire data skills while majoring in another field. Two-year institutions are starting to introduce associate's degrees and certificates in data science to prepare students to transfer to 4-year

---

[1] For information about community efforts toward more transparent data-driven decision making for social good, see http://datafordemocracy.org, accessed March 12, 2018.

programs or to give them skills to compete in the workforce. Summer programs enable undergraduate students to build up data science skills rapidly. Boot camps and intensive training programs that aim to refresh or retool postgraduate students with the skills required of the growing data science workforce are now appearing. Massive open online courses in data science are proliferating and serve as stand-alone points of entry for all kinds of students (and flexible opportunities for professional development for instructors).

These pioneering examples of programs show what is possible, but there are significant challenges to developing data science programs more broadly and pervasively. The popularity of data science courses and programs will affect the entire academic institution by influencing enrollment, budgets, classroom allocation, computing resources, and scheduling. Institutions may need to consider how to create incentives for faculty in multiple departments and fields to collaborate to develop and deliver curricula that best meet students' needs. Today, there is a shortage of faculty in this rapidly evolving area. Enlisting and training existing faculty will be essential in the short term, and developing new faculty will be important in the long term. These challenges, among others, will need to be addressed to ensure the success of undergraduate data science students.

Recommendation 5.1: Because these are early days for undergraduate data science education, academic institutions should be prepared to evolve programs over time. They should create and maintain the flexibility and incentives to facilitate the sharing of courses, materials, and faculty among departments and programs.

The evolution of data science programs will be affected by a broad range of factors, including their initial home and structure, the needs and interests of students, and institutional culture. Although new programs could be launched by combining existing courses and materials, over time new classes and materials will need to be developed. Institutions will need to think through the pathways students are taking into data science and how to create bridges and remove barriers. Academic and career advising will be vital parts of data science programs; the advising programs will themselves need to evolve as the field and the market for graduates mature.

Data science itself provides tools to continuously evaluate and improve data science education. Evaluation should include assessment of student learning and assessment of how well a program is meeting the needs of the market it aims to serve. Evaluation can be used to shape a program at a given institution, showing what is working and where improvement is needed. It can also be used comparatively to detect approaches, classes, or curricula that may be of value to other campuses or contexts.

> **Recommendation 5.3: Academic institutions should ensure that programs are continuously evaluated and should work together to develop professional approaches to evaluation.** This should include developing and sharing measurement and evaluation frameworks, data sets, and a culture of evolution guided by high-quality evaluation. Efforts should be made to establish relationships with sector-specific professional societies to help align education evaluation with market impacts.

Much of the necessary data for evaluation could come from institutions' administrative records. These records, used in conjunction with other data sources such as economic information and survey data, could enable effective transformation and generalization of programs and might even inform a cohesive national approach to undergraduate data science education.

In many fields, professional societies play a role in creating and nurturing community, in facilitating the sharing of resources and results, and in convening groups to set standards or determine best practices. Such capabilities are valuable to data science as well. However, it may be difficult for a single existing society to represent all the interests of the data science community. A structured collaboration of existing professional societies might work better, with potential development of subsocieties devoted to data science elements in any of many preexisting societies.

> **Recommendation 5.4: Existing professional societies should coordinate to enable regular convening sessions on data science among their members.** Peer review and discussion are essential to share ideas, best practices, and data.

Conferences, workshops, training sessions, and other networking opportunities would benefit the joint communities. Other opportunities for the collaborating societies would be collecting materials; convening discussions around critical topics such as curriculum, evaluation, and ensuring broad participation; and potentially creating publication venues for the broad community. *As data science continues to evolve, it is essential that academic institutions and other stakeholders take steps to prepare students for a data-enabled world. The time to act is now.*

1

Introduction

In the past decade, the world has been transformed by the rapidly evolving field of data science. This new science, which is already revolutionizing business, science, and society, builds on an array of technological developments, including the widespread use of smartphones and rapid technological progress in computing and communications. Massive investments have gone into building out wireless infrastructure and data centers (the cloud) and into leveraging such facilities. New methods have been developed to connect and understand the data being generated.[1]

In this new landscape, all individuals constantly generate data about their whereabouts, habits, and preferences. All parts of commerce—browsing, ordering, shipping, inventory, manufacturing, advertising—have gained a digital footprint. Social network sites have illuminated relationships among billions of individuals, and tweets and posts have made global-scale communication patterns instantly visible. Governmental bodies have digitized and given public access to vast corpora of data and documents. Most of recorded history and literature have become digitized and accessible for algorithmic analysis. Electronic health records have allowed medical analyses across populations and time, while genomic sequencing has brought individualized treatment to the cellular level. Design and synthesis of pharmaceuticals, materials, and

---

[1] Science and engineering have provided many notable examples of digital transformations in the previous decade, foreshadowing the large public transformation now taking place. Examples from the 1995-2005 time frame include virtual observatories (see, e.g., Szalay and Gray, 2001; NSF, 2018) and advanced computational methods (see, e.g., Berman, Fox, and Hey, 2003).

chemicals have become computational. The volume of data being collected automatically—and the processing of such data—has soared. New data-driven services have arisen (e.g., navigation apps, ride-hailing apps, and voice-driven assistants), exploiting this new data-driven environment and convincing the public of the power and elegance of the data-driven paradigm. Several of the highest market capitalization companies have been heavily involved in digital transformation, displacing oil and car companies that had been market leaders for decades.

These emblematic advances signal more extensive and widespread transformations to come. The smartphone, mobility, genomics, and cloud "revolutions" are in fact only at their inception as technologists find ways to leverage them ever further. The increased use of Internet-connected home thermostats and fitness wristbands have marked the beginning of the Internet-of-Things era, in which people are surrounded by an environment that is instrumented, communicative, and responsive. Meanwhile, rapid advances in machine learning are enabling new applications.

This year's entering undergraduates, who may be in the workforce until roughly 2075, will face an employment landscape transformed by these developments. The data-driven era will spawn many new occupational niches based on the massive opportunities presented by new kinds and volumes of data even as it supplants traditional occupational categories.

Today, the term "data scientist" typically describes a knowledge worker who uses the complex and massive data resources characteristic of this new era. However, data science is a broader concept involving principles for data collection, storage, integration, analysis, inference, communication, and ethics appropriate for this new data-driven era. Several industries and academic disciplines have perceived that a new field of data science is emerging out of several established fields, including information technology, computer science, statistics, mathematics, operations management, and business analytics. However, core data science concepts involving the aforementioned principles are not being conveyed by mainstream training in any one field because data science is not reducible to any of the preexisting fields. Data scientists of the future will need to be educated in the full scope of data science principles.

There are many reports that industry finds itself constrained by today's relatively small supply of well-trained data science talent, and hiring demand for data scientists has begun to increase rapidly; some projections forecast that approximately 2.7 million new data science positions will be available by 2020 (Columbus, 2017). Not only is the lack of data science talent an issue, but so too is students' lack of understanding about what a data scientist is and what types of tasks such an individual might perform.

It is imperative that educators, administrators, and students begin today to consider how to best prepare for and keep pace with this data-driven era of tomorrow. Undergraduate teaching, in particular, offers a critical link in providing more data science exposure to students and expanding the supply of data science talent. Many distinct data science roles will exist in the future workplace; both specialists and broad users with different levels of knowledge and different skill sets will be in high demand.

Understandings of and applications for data science vary among professionals, within academic institutions, and throughout the broader world. One common observation is that data science is now essential in many academic fields (Hey, Tansley, and Tolle, 2009) and can be both pervasive in and yet distinct from other disciplines. For example, data science techniques and tools may be applied commonly across a variety of disciplines, including those in the sciences and in the humanities. However, what gives data science its unique identity is that it draws on individual skills and concepts from a wide spectrum of disciplines that may not always overlap with one another—a truly multidisciplinary field. As discussions continue regarding the distinctions among data science, computer science, statistics, and other fields, many U.S. academic institutions are considering how to best deliver data science education and thus better prepare graduates for the data-driven era that lies ahead of them.

The need for data science instruction is broad and extends to a wide range of students from varied programs. Depending on the students' levels of interest and career goals, as well as institutional goals and resources, one can envision a variety of models for data science instruction, including discipline-centered data science courses offered by specific academic departments focusing narrowly on the skills needed by that department's majors, large introductory data science courses serving the campus-wide student body, highly structured course sequences within a formal data science major, online courses, boot camps, and other innovative approaches. To achieve this vision, data science education and practice demand a level of collaboration not necessarily seen in other fields, new approaches to evaluating educational outcomes, and a constant eye toward refining and evolving the undergraduate experience as this field continues to advance. Stand-alone data science departments may emerge naturally on some campuses when the level of collaboration surpasses the bandwidth of currently established departments or when the student demand increases greatly. However, developing stand-alone departments is not the only means of delivering data science education effectively nor may it be appropriate in all settings—equipping students with data science skills can be done through a variety of pathways, as will be discussed in this report.

A LOOK TO THE FUTURE

Imagine it is now 2040. Students born in 2018 are graduating from college. It is more than 30 years since billions of autonomous sensors and devices started continuously delivering data to cloud-based databases, which record the states and activities of vehicles, buildings, customers, patients, and citizens. Many other data-driven changes that were difficult to foresee have become pervasive and important. Thus, it is not farfetched to expect academic institutions to envision the data-driven world of 2040 as they shape the future undergraduate experience.

In the ideal case for the future evolution of data science, all private industries and public agencies would use data confidently and efficiently to operate fairly without gender or racial bias. Data science jobs would be plentiful. While some of these data science jobs would require vocational education, other data science subspecialties would require certificates, associate's degrees, and bachelor's degrees. Efforts would have been undertaken to distribute the workforce equitably over rural, urban, and suburban regions; socioeconomic strata; and ethnic identities. The importance of data skills would be appreciated in all high schools, and the vast majority of high school graduates would have a basic understanding of data science. Data science methods would be used by data science programs to continuously evolve to meet the needs of their students.

Data scientists' work would be varied, and different skill mixes would be needed for different data science positions. Some of these individuals would have been trained in particular fields but have learned data science along the way. Others would have explicit degrees in data science. For those who need a degree in data science for their work, there will likely be many options. They might earn those degrees remotely, on-site, or in combination. They might learn through a mixture of interactive web applications and augmented reality simulations, interactions with fellow learners and multidisciplinary faculty, and immersive industry apprenticeships. Students in 2- and 4-year institutions would be exposed to important concepts through a range of motivating applications. Humanities, social sciences, and professional education (e.g., music, art, and architecture) would be taught for enrichment, for building cross-disciplinary communication skills, and as contexts in which to provide examples of different types of data. Ethical data concepts such as privacy, justice, fairness, and reproducibility would be taught continuously in safe spaces where students learn from their mistakes without penalty and without harm to others. Faculty would use data science to continuously monitor their students' progress and to adapt their curriculum to ensure student competency, confidence, and well-being with respect to the needs of industry, government, and society.

The committee's vision for the world of 2040 has many debatable elements—whether the transformations just described will actually go nearly as far as depicted or whether this mostly utopian vision will develop dystopian elements. This much is not debatable: the undergraduate instructional framework will need to transform if it is to support the transition from the world of 2018 to the likely world of 2040. This report outlines some considerations and approaches for academic institutions and others in the broader data science communities to help guide this transformation, but it is not intended to be a final word on undergraduate data science education. This vision needs to be continually evolved and refined as the field matures.

REPORT OVERVIEW

In Chapter 2, the committee considers *what* data science professionals will need to know. Because expectations and tasks for data scientists will vary across industries and over time, it is important to consider the skill sets, learning outcomes, and ethical considerations best suited for individual undergraduate students to be successful in their future careers. In Chapter 3, the committee lays the groundwork for exploring *how* these data science students can be educated and thus well prepared. Using data from existing data science education programs, the committee discusses the successes and challenges associated with implementing and delivering 2- and 4-year undergraduate programs and classes, alternative courses, and interdisciplinary approaches in an effort to guide individual institutions to follow the pathways that simultaneously align with their missions and meet the varied needs of the field of data science. In Chapter 4, the committee describes a number of challenges that arise in creating a new data science program. Acknowledging that the field of data science and the content of data science education will continue to change rapidly, the committee considers how to evolve from current to future data science education and practice in Chapter 5. The committee evaluates strategies to refine educational and administrative infrastructure, create professional development opportunities, and draw on professional societies. In Chapter 6, the committee offers a summary of its findings and recommendations that appeared throughout Chapters 2 to 5.

REFERENCES

Berman, F., G. Fox, and A.J.G. Hey, eds. 2003. *Grid Computing: Making the Global Infrastructure a Reality*. West Sussex, UK: Wiley.

Columbus, L. 2017. IBM predicts demand for data scientists will soar 28% by 2020. *Forbes*, May 13.

Hey, T., S. Tansley, and K. Tolle, eds. 2009. *The Fourth Paradigm: Data-Intensive Scientific Discovery*. Redmond, Wash.: Microsoft Research.

NSF (National Science Foundation). 2018. The Observatory [National Ecological Observatory Network]: History. http://www.neonscience.org/observatory/history. Accessed February 6, 2018.

Szalay, A., and J. Gray. 2001. The world-wide telescope. *Science* 293(5537):2037-2040.

2

Knowledge for Data Scientists

Over the past decade, data science has emerged out of a variety of widespread developments (as discussed in Chapter 1), and companies, academic institutions, and governments are striving to hire data scientists while transforming their practices (BHEF and PwC, 2017; Ernst and Young, 2017). There are many instances of academic data science. Still, "data science" is not yet fully defined as an academic subject; the central tenets, concepts, knowledge, skills, and ethics powering this emerging discipline remain points of active discussion and continue to evolve. A new generation of tool developers and tool users will require the ability to understand data, to make good judgments about and good decisions with data, and to use data analysis tools responsibly and effectively (referred to as "data acumen" throughout this report). Developers and users draw from computing, mathematics, statistics, and other fields and application domains. Educators and administrators are beginning to reimagine course content, delivery, and enrollment at the undergraduate level to best prepare students to operate in this new discipline.

New and greater volumes of information, along with its variety and velocity, compound long-standing challenges of data analysis—and raise new ones. The ability to measure, understand, and react to large quantities of complex data can shape scientific discovery, social interaction, political interactions and institutions, economic practice, public health, and many other areas. Data science workflows not only consume data, but they also produce data—such as intermediate data sets, statistics, and other by-products such as visualization—that need to be understood.

Although the definition of data science is evolving, it centers on the notion of multidisciplinary and interdisciplinary approaches to extracting knowledge or insights from large quantities of complex data for use in a broad range of applications. Data science is about synthesizing the most relevant parts of the foundational disciplines to solve particular classes of problems or applications that are newly enabled because the volume and variety of data available are expanding swiftly, data are available more immediately, and decisions based on data are increasingly automated and in real time. Data scientists often work at the interface of disciplines and can help develop new approaches to address problems in these areas.

Data science applications have varying levels of risk. For example, recommender systems that suggest purchases within an online shopping platform or select advertisements for website visitors are relatively low risk. Although provider sales may be affected if undesirable products are recommended and users may be dissatisfied with their purchases, the overall impact of poor retail recommender systems to individuals and society is generally low. Still, the recommendations can influence the behavior of large segments of a population and are often coupled with a just-in-time supply chain, which aims to forecast consumer demand given available data and optimize production and shipping of goods. In this case, the systems can have substantial impact, especially if they result in a shortage of necessary items, such as food and medicine, owing to natural disaster or unanticipated interactions with other external factors. But increasingly, as similar data-driven algorithms are used to recommend sentencing or release of criminals, guide testing or treatment of patients, plan urban development, draw political boundaries, allocate funds, and inform other critical public policy decisions, impacts on individuals and society can be profound.

While new volumes and types of information can make analyses more accurate than past methods that relied on sparse surveys with lower than desired survey frequency, response rates, and sample sizes, they still have limitations. Weaknesses in data quality and data analysis might have a wide range of negative policy effects: problems might be misunderstood in their causes and scale; a program that a family depends on might get insufficient funding; or a policy might be enacted that has unintended consequences for large segments of the population.

Thus, it will be important that data are collected and analyzed appropriately and that there are clear principles guiding the use of data for human good. Furthermore, the complexity of the analyses and the increasing dependency on data across all fields of human endeavor will drive demand for "smarter" tools and best practices for data science that minimize mistakes in interpretation.

Data science is not just the practice of analyzing a certain data set about a particular question. It often results in the creation of processes that continuously take in new data, often from many sources, and generate refined distillations of those data, which in turn become sources for new inquiries, questions, and analyses. The products of the data scientist—including data, code, visualizations, and recommendations—often take on a life of their own far beyond the initial question that gave rise to their creation. In this way, data science takes on aspects associated with engineering—namely, the creation of infrastructures that undergird society and must safely withstand unanticipated changes in demand and use.

Academic institutions, companies, and governments recognize these shifts and are rapidly embracing a vision of an emerging discipline of data science that is unique yet builds on knowledge from existing disciplines (NRC, 2014). Generally, each academic discipline recognizes that its viewpoint alone is insufficient to encompass all of data science.[1] Advances in the power and usability of data science computing tools have made it possible for even inexperienced people to conduct complex analyses over enormous data sets without really understanding the possible artifacts and biases that may be lurking in the data or the reliability of the results and interpretations. Machine learning models may achieve superhuman performance on challenging machine vision tasks yet may employ biased or unfair interpretations of the data (Jordan, 2013). Application domains (e.g., business, medicine, natural science, social sciences, or engineering) are developing and adapting machine learning and deep learning techniques to solve specific research questions. These techniques can be more effective than previously used methods but may lack mathematical or statistical rigor or computational scalability. Increasingly, domains in the humanities, such as philosophy, rhetoric, history, and literary studies, embrace elements of data science while issues of algorithmic bias present moral and ethical questions.

Data scientists have the potential to help address critical real-world challenges. Just a few examples are listed here:

- *Enabling more accurate diagnosis of melanomas through better analysis of images.* Deep learning techniques have been applied to detect melanoma, the deadliest form of skin cancer. These methods improve

---

[1] The American Statistical Association and the Computing Research Association have both released formal statements to this effect (ASA, 2015; CRA, 2016). The Institute of Electrical and Electronics Engineers (IEEE) has introduced numerous data science conferences associated with its various special interest groups. National position statements around information management and operations research are less well defined. The popular press is full of comparisons of data science and business analytics or business intelligence; none assert that the latter two subsume data science.

the analysis of tissue images, promising a more accurate diagnosis than traditional techniques (Codella et al., 2017).
- *Enhancing business decisions.* Business analytics can assist entrepreneurs and company executives in making timely decisions based on market trends. This can be coupled with analysis of online social media information to respond directly to consumer demands or create a more personalized advertising experience (Chen, Chiang, and Storey, 2012).
- *Helping aid organizations to respond faster.* Data science and analytics are used to assist aid organizations to respond more quickly in times of need, such as when the Swedish Migration Board used data science to make predictions about and determine national implications of emigration trends (Pratt, 2016).
- *Developing "smart cities."* Cities around the world, such as London, Rio de Janeiro, and New York, collect real-time data from a variety of sources, such as public transportation, traffic cameras, environmental sensors for parameters such as temperature and humidity, and social media interactions regarding local issues. The data can then be processed, analyzed, and utilized to improve city efficiency and cost-effectiveness as well as resident well-being (Kitchin, 2014).

However, there are also many instances of high-impact and high-profile data science research that has resulted in flawed or inaccurate findings, as well as ethical and legal quandaries. A few examples are listed here:

- *Inaccurate predictions of flu trends.* In 2013, Google Flu Trends overpredicted true influenza-related doctors' visits as determined by the Centers for Disease Control and Prevention. This has been primarily attributed to overreliance on outdated models (Butler, 2013).
- *Release of personally identifiable data.* The abundance of data available on individuals from companies and social media can present ethical dilemmas to researchers in terms of privacy, scalability of results, and subject participation agreement. For instance, a 2013 study linking numerous Twitter users to sensitive information from their financial institutions prompted discussions of when researchers should be required to obtain written consent when using nominally publicly accessible information (Danyllo et al., 2013).
- *Biases in predictive policing.* There is much debate over the use and appropriateness of predictive policing—the use of data science by law enforcement to predict crime before it occurs. There is no consensus yet on the effectiveness of this methodology, and civil liberties groups argue that the data used to develop (i.e., train) the models are inherently biased (Hvistendahl, 2016).

- *Surveillance of citizens.* China is deploying facial recognition technologies as well as other data science approaches to track individuals and influence behavior. The national goal is to link these surveillance systems by 2020 to "implement a national 'social credit' system that would assign every citizen a rating based on how they behave at work, in public venues, and in their financial dealings" (Chin and Lin, 2017).

Data science is currently being applied in many organizations within industry, academia, and government, often by self-taught practitioners. There are indications of strong demand in a variety of domains for graduates with data science skills. A recent study by IBM found more than 2.3 million data science and analytics job listings in 2015, and both job openings and job demand are projected to grow significantly by 2020 (Columbus, 2017). Three-fifths of the data science and analytics jobs today are in the finance and insurance, professional services, and information technology sectors, but the manufacturing, health care, and retail sectors also are hiring significant numbers of data scientists (Markow et al., 2017). The IBM study also shows that it takes significant time to find and hire staff with the right mix of skills and experience (Columbus, 2017). Since many employers are themselves new to the use of data science, they may not be able to provide training and therefore may prefer to hire individuals who already have appropriate classwork and hands-on experience. More generally, a poll conducted by Gallup for the Business-Higher Education Forum revealed that 69 percent of employers expect candidates with data science and analytics skills to get preference for jobs in their organizations by 2021 (BHEF and PwC, 2017).

Current data science courses, programs, and degrees are highly variable in part because emerging educational approaches start from different institutional contexts, aim to reach students in different communities, address different challenges, and achieve different goals. This variation makes it challenging to lay out a single vision for data science education in the future that would apply to all institutions of higher learning, but it also allows data science to be customized and to reach broader populations than other similar fields have done in the past. Moreover, the continual emergence of new data sources and new analytical tools make this an extremely fluid environment, where the courses that are taught today might be organized around concepts and practices that are supplanted in the near future. Any data science program will have to take this into account, and this complicates discussions about how to define and structure the field.

However, important foundational data science skills are highlighted in this chapter and may serve as a platform for any practicing data scientist. The themes described in this chapter underlie data science education, but they are not necessarily novel challenges or even unique to data science. The lessons learned from other disciplines can help pave the way to ensuring the success of data science education.

DATA SCIENTISTS OF TODAY AND TOMORROW

As was discussed in the previous section, there is a current shortage of workers with data science skills. The day-to-day work and thus educational needs of the different types of data scientists are highly differentiated. This section of the report will map the educational needs with the roles that students will be expected to perform in the workplace and the skills needed to prepare students for graduate studies and research careers in many fields of inquiry (NRC, 2013).

Data science roles vary across government, industry, and academia and will continue to evolve in the future. As with other complex fields of study, there is both differentiation and overlap in these roles. The breadth and depth of data science roles underscore the complexity that employers face in the identification of qualified candidates for their job postings and the challenges that academic institutions face in preparing their students for these emerging roles. Some current areas of focus for data scientists include the following:

- *Computing hardware and software platforms for data science.* Data scientists who manage the platforms on which data science models are created focus on understanding and maintaining a computing environment that meets the demands for big data, fast (sometimes real time or near real time) model generation, and data interrogation—up to and including the demands of real-time data collection (i.e., streaming) and complex data visualizations. A significant challenge of this job is remaining current on the latest computing hardware and software. Unlike system administrators, who need to understand only one or two computer systems, these data scientists create environments for data science modelers and analysts that can be used across a range of computing platforms. This requires that they understand the changing programming languages used for data science, the supporting libraries, and the many types of data storage systems, as well as how to keep all of these components operational and secure. Because of the rapid rate of change in this area, educational training needs to focus on key topics such as database maintenance, security, programming hardware, and operating systems. A certain level of proficiency in these skills could be developed in a 2-year associate's degree program, upon completion of which graduates will be able to manage changing computing systems and keep pace with the ever-growing computational needs of machine learning and data science model development and workflows. However, additional depth may be required to equip professionals to keep pace with rapidly advancing technology, perform capacity planning and availability assessments, and deploy solutions that are reliable and scalable.

- *Data storage and access.* Data scientists who focus on managing data storage solutions as well as extracting, transforming, and loading data for modeling should have the ability to manage exceptionally large data sets from a variety of heterogeneous data sources and in batch or streaming form, and to assess the predictive value of these data sources. A strong knowledge of both databases and streamed analytical processing is key to this role. These data scientists need to understand the data science workflow, to document data quality problems, and to select appropriate methods of interpolation—even, in some cases, creating data models to clean and reduce errors in downstream model development performance. Some domain knowledge is likely needed (e.g., to understand data quality issues and how to best mange the data). The education needed for this role varies, and skill sets could be developed at both 2- and 4-year institutions (keeping in mind that a data science team would likely need to represent additional important skills, such as computing, continuous cross-validation, and adoption of new modeling techniques or frameworks).

- *Statistical modeling and machine learning.* Experts in statistical modeling and machine learning interface with stakeholders to capture requirements and develop the scope of work for data science projects, undertake the data science analysis cycle, and typically bridge the gaps among more narrowly focused data science roles. Written and oral communication skills are essential for this position, as is experience with coordinating teams. Often these data scientists require considerable domain expertise in the field for which the data science models are being developed. For example, an individual developing a model for clinical trial analysis for drug development would need to have a significant understanding of pharmacology and clinical data collection. Although data science skills required for this role are broad, the disciplinary knowledge is highly specific. Owing to the breadth and complexity of this position, a 4-year undergraduate program may be required to develop the level of proficiency necessary for success. However, even 4-year programs are unlikely to develop sufficient knowledge in a domain through exposure to a small number of courses (i.e., a second

major, co-major, or minor may be necessary). All data scientists need to acquire domain knowledge, but it is particularly important for this role.
- *Data visualization.* Ideally, data visualization experts combine development and design skills with the ability to understand the meaning of the underlying analyses. These data scientists are adept at visual storytelling with data. They can examine large data sets and create clear, efficient, compelling online layouts, images, dashboards, and interactive features that can stand on their own or complement narrative text. At their core, they are effective translators between technical and statistical specialists and superior communicators with multiple nontechnical audiences. They are well versed in the key elements of effective graphical displays as well as the pitfalls of misrepresenting data and results. These data scientists combine knowledge of statistical analysis tools, libraries, and frameworks to complement a foundation in computational, statistical, and data management methods. They are well prepared to adjust quickly as new standards and tools become available. They can design for multiple formats and platforms and be grounded in user experience insights. They understand application programming interfaces—how to parse them and, ideally, how to build them—and are closely aligned with the data management functions performed by others on a team. Both 2- and 4-year programs can help prepare students for this role.
- *Business analysis.* A growing number of positions involves making sense of and communicating about data without necessarily relying on programming skills. These jobs are built around assembling and presenting data to inform a decision-making process. These data scientists are common in many business areas, have expertise in various domains, and can utilize skills developed in both 2- and 4-year programs.

There are many other types of data scientists today, and their roles will continue to change and expand in the future. Beyond the differences among them, there is considerable variance in the lower-order and higher-order knowledge and skills that some data science jobs require. There are also many commonalities among the varied types of data scientists. All data scientists need to learn how to tackle questions with real data. It is insufficient for them to be handed a "canned" data set and be told to analyze it using the methods that they are studying. Such an approach will not necessarily prepare them to solve more realistic and complex problems taken out of context, especially those involving large, unstructured data. Instead, they need repeated practice with the entire cycle beginning with ill-posed questions and "messy" data.[2]

An effective data science workflow involves formulating good questions, considering whether available data are appropriate for addressing a problem, choosing from a set of different tools, undertaking analyses in a reproducible manner, assessing analytic methods, drawing appropriate conclusions, and communicating results. Students need practice applying a unified approach to problem solving with data. Such an integrated approach needs to be introduced in their first courses and remain a consistent theme in subsequent courses. Students need to see that data science is not simply a collection of varied tools (or methods), but rather a general approach to problem solving. Many of the emergent data science programs at every academic level encourage students to assume that they will benefit from continuing professional education throughout their careers. All require that graduates have the capability to identify problems to be solved with data, determine and implement solutions, assess results, and communicate results and findings (UC Santa Cruz, 2018).

Finding 2.1: Data scientists today draw largely from extensions of the "analyst" of years past trained in traditional disciplines. As data science becomes an integral part of many industries and enriches research and development, there will be an increased demand for more holistic and more nuanced data science roles.

Finding 2.2: Data science programs that strive to meet the needs of their students will likely evolve to emphasize certain skills and capabilities. This will result in programs that prepare different types of data scientists.

Recommendation 2.1: Academic institutions should embrace data science as a vital new field that requires specifically tailored instruction delivered through majors and minors in data science as well as the development of a cadre of faculty equipped to teach in this new field.

Recommendation 2.2: Academic institutions should provide and evolve a range of educational pathways to prepare students for an array of data science roles in the workplace.

---

[2] A description of the importance of the multistep scientific process and how it relates to data analysis can be found in the *Curriculum Guidelines for Undergraduate Programs in Statistical Science* (ASA, 2014).

DATA ACUMEN

Data science is a complex activity that requires specific skills, such as coding in advanced computer languages, and less well-defined but equally crucial skills, including the ability to do the following:

- Combine many existing programs or codes into a "workflow" that will accomplish some important task;
- "Ingest," "clean," and then "wrangle" data into reliable and useful forms;
- Think about how a data processing workflow might be affected by data issues;
- Question the formulation and establishment of sound analytical methods; and
- Communicate effectively about properties of computer codes, task workflows, databases, and data issues.

Aspiring data scientists need to develop these skills in order to avoid conducting flawed or incomplete analyses.

In short, getting a useful answer from data requires many skills that are often not fully developed on their own in traditional mathematics, statistics, and computer science courses—although such fields certainly come closest today to providing mastery of the desired skill set. Donoho (2017) noted the need for data scientists who can face "essential questions of a lasting nature and [use] scientifically rigorous techniques to attack those questions."

Students also need to learn how to ensure that outcomes are valid—extracting the right insights and having confidence that, start to finish, what one says is true, within some margins of error. Repeated exposure to the data science life cycle (i.e., posing a question; collecting, cleaning, and storing data; developing tools and algorithms; performing exploratory analysis and visualization; making inferences and predictions; making decisions; and communicating results) is needed to help hone the skills required to assess the data at hand, extract meaning from them, and communicate those findings to nonexperts. Students also need to consider the provenance of the data used.

Building on the work of De Veaux et al. (2017), the committee puts forth the following key concept areas for data science: mathematical foundations, computational foundations, statistical foundations, data management and curation, data description and visualization, data modeling and assessment, workflow and reproducibility, communication and teamwork, domain-specific considerations, and ethical problem solving.

Experience and facility in these and other areas are essential to building what this committee defines as "data acumen." Some exposure to key high-level topics is needed by all students, while other students will require additional exposure or extended work to develop expertise. The process of starting students down the path toward data acumen is a chief objective of data science education.

Finding 2.3: A critical task in the education of future data scientists is to instill data acumen. This requires exposure to key concepts in data science, real-world data and problems that can reinforce the limitations of tools, and ethical considerations that permeate many applications. Key concepts involved in developing data acumen include the following:

- Mathematical foundations,
- Computational foundations,
- Statistical foundations,
- Data management and curation,
- Data description and visualization,
- Data modeling and assessment,
- Workflow and reproducibility,
- Communication and teamwork,
- Domain-specific considerations, and
- Ethical problem solving.

Recommendation 2.3: To prepare their graduates for this new data-driven era, academic institutions should encourage the development of a basic understanding of data science in all undergraduates.

Mathematical Foundations

Mathematics is essential for data science; however, how much and what types of mathematics are needed vary. Data scientists need to know how to test hypotheses and determine why they do or do not align to real-world problems. They need to be capable of assessing their data science models, determining when these models fail and how to make corrections that lead to scientific discovery. Tools (e.g., Wolfram Alpha) can be utilized and combined to produce an outcome (e.g., simulation or visualization) that reinforces data scientists' computational and statistical knowledge without demanding the study of calculus in full detail (see Hardin and Horton, 2017).

New, more flexible pathways to help establish a mathematical foundation for data science are being developed. The University of Texas at Austin Dana Center's Mathematics Pathways[3] is one such program designed to increase opportunities for students across the nation through mathematics and statistics education. This program instills confidence, advocates for degree or certificate completion, and provides students with the skills and tools to apply mathematical and quantitative reasoning at home and in the workplace. The development of additional pathways to help students develop mathematical foundations would be beneficial for the field of data science.

Key mathematical concepts/skills that would be important for all students in their data science programs and critical for their success in the workforce are the following:

- Set theory and basic logic,
- Multivariate thinking via functions and graphical displays,
- Basic probability theory and randomness,
- Matrices and basic linear algebra,
- Networks and graph theory, and
- Optimization.

Some data scientists and programs require a deeper understanding of mathematical underpinnings. This might include the following:

- Partial derivatives (to understand interactions in a model),
- Advanced linear algebra (i.e., properties of matrices, eigenvalues, decompositions),
- "Big O" notation and analysis of algorithms, and
- Numerical methods (e.g., approximation and interpolation).

While linear algebra and optimization may be particularly helpful in data science, the traditional mathematics curriculum has many courses that precede multivariate calculus and linear algebra. It may be the case that institutions need to develop a "math for data science" class[4] to build these foundations without requiring multiple semesters of coursework. This could potentially serve as an accelerated course in relevant mathematical approaches for data science and possibly replace further coursework for some students.

---

[3] The website for the Dana Center's Mathematics Pathways is http://www.utdanacenter.org/higher-education/dcmp/, accessed January 18, 2018.

[4] See Hardin and Horton (2017) for one suggested approach.

Computational Foundations

Working with data requires extensive computing skills. Data science graduates need to be proficient in many of the foundational software skills and the associated algorithmic, computational problem-solving skills associated with the discipline of computer science. A data science student needs to be prepared to work with data as they are commonly found in the workplace and research laboratories. Accessing and organizing data in databases, scraping data from websites, processing text into data that can be analyzed, ensuring secure data storage, and protecting confidentiality all require extensive computing skills. Computational problem-solving skills recur throughout the data scientist's workflow. As Wing (2006, p. 34) noted, "Thinking like a computer scientist means more than being able to program a computer. It requires thinking at multiple levels of abstraction."

To be prepared for careers in data science, students also need facility with professional statistical analysis software packages and an understanding of the computational and algorithmic problem-solving principles that underlie these packages.

It is also important for data science students to be aware of the state of the art of information technology and for faculty to educate these students so that their knowledge will continue to evolve accordingly. Students will also benefit from instruction in aspects of data structures, object-oriented programs, and workflow (i.e., aspects of a broader set of project management skills). The first pedagogical approach to achieving this understanding is to teach students how to think about algorithms. Students will need further skill development to be able to deepen their understanding of abstraction and be able to learn new data technologies. It is more important for students to learn how to follow the information technology frontier than to master the details of today's architecture.

While it would be ideal for all data scientists to have extensive coursework in computer science, new pathways may be needed to establish appropriate depth in algorithmic thinking and abstraction in a streamlined manner. This might include the following:

- Basic abstractions,
- Algorithmic thinking,
- Programming concepts,
- Data structures, and
- Simulations.

Statistical Foundations

All data scientists need to understand basic statistical concepts, practice, and theory. According to De Veaux et al. (2017, p. 20), "Students should understand the basic statistical concepts of data analysis, data collection, modeling, and inference. A sound knowledge of basic theoretical foundations will help inform their analyses and the limits to their models. Successful graduates will be able to apply statistical knowledge and computational skills to formulate problems, plan data collection campaigns or identify and gather relevant existing data, and then analyze the data to provide insights."

To avoid drawing invalid or incorrect conclusions, data science students need to understand the concept of inference, including sampling and nonsampling errors. Owing to the nature of observational data as found artifacts (which may represent a nonrandom selection or include confounding factors), it is important for students to study confounding and causal inference early to make sense of the data around them. As a specific example, having 30 million credit card records can help identify a number of relationships in the observed data (e.g., people who shop at a particular retailer tend to exceed a certain income threshold), but those relationships will not necessarily hold in the next set of records. In addition, other measured or unmeasured factors may be important in determining causal conclusions (e.g., students could wrongly conclude that use of sunscreen is associated with skin cancer if the amount of sun exposure is not controlled for in an analysis).

As for the previous areas, work is needed to identify approaches to build a strong foundation in statistics. The American Statistical Association (ASA) guidelines for undergraduate programs in statistics (ASA, 2014) discuss important considerations for educating students in statistical practice, as do the "Curriculum Guidelines for Undergraduate Programs in Data Science," which were endorsed by the ASA (De Veaux et al., 2017). Data science students need to know about randomized trials (commonly used in businesses running A/B comparisons) but need to quickly move to approaches that are applicable for nonrandomized studies. They need repeated practice with the whole data science life cycle.

Important statistical foundations might include the following:

- Variability, uncertainty, sampling error, and inference;
- Multivariate thinking;
- Nonsampling error, design, experiments (e.g., A/B testing), biases, confounding, and causal inference;
- Exploratory data analysis;
- Statistical modeling and model assessment; and
- Simulations and experiments.

Data Management and Curation

At the heart of data science is the storage, preparation, and accessing of data. It is often said that a typical data analysis project is more than 70 percent data cleaning, merging, and marshaling. With the advent of large public databases with data of all kinds ranging from governmental to genomic, there has never been a better time to teach students about the many aspects of data management. The students can directly experience the many forms in which data can be found today, from spreadsheets and text files to relational and nonrelational databases.

One way in which data scientists succeed is by providing others a very clear understanding of the details of the data that went into a project, possibly also making the data, or their derivatives, available to others. Throughout their coursework, students need to become facile with data of different types (e.g., relational, text, images).

Key data management and curation concepts/skills that would be important for all students in their data science programs and critical for their success in the workforce are the following:

- Data provenance;
- Data preparation, especially data cleansing and data transformation;
- Data management (of a variety of data types);
- Record retention policies;
- Data subject privacy;
- Missing and conflicting data; and
- Modern databases.

Data Description and Visualization

Many data scientists create value by creating "dashboards" that display some basic statistics and visualizations to monitor the contents of an evolving database or stream. In this way, they provide situational awareness for decision makers. Students who might be creators or users of such dashboards need to learn about traditional descriptive statistics for developing a feel of what is in a data set as well as about traditional graphics such as scatter and time-series plots with decorations and modifications. This will help prepare them to present data in a clear and compelling fashion. After learning how to make basic displays, students then need to be taught how to use simple graphics to check data for artifacts, snafus, and inconsistencies. Then they can start to undertake exploratory data analysis.

Data visualization is at the core of data science insight extraction, communication with others, and quality assurance. A key challenge for data scientists is to be able to tell a story with data and translate key aspects of the data science life cycle and outcomes of efforts to both users and leaders. It is crucial that data visualization training goes hand-in-hand with communication training, as a well-chosen graph can efficiently convey to others some important feature of a data set that might otherwise be very difficult to capture in words. Such visual displays help to avoid a "garbage in, garbage out" situation where important outliers or incorrectly coded data lead to misleading conclusions.

Key data description and visualization concepts/skills that would be important for all students in their data science programs and critical for their success in the workforce are the following:

- Data consistency checking,
- Exploratory data analysis,
- Grammar of graphics,
- Attractive and sound static and dynamic visualizations, and
- Dashboards.

Data Modeling and Assessment

Data scientists have a rich and growing set of models and methods at their disposal. The challenge is how to identify which models are most appropriate for a given setting and assess whether the assumptions and conditions needed to apply that method are tenable.

Key data modeling and assessment concepts/skills that would be important for all students in their data science programs and critical for their success in the workforce are the following:

- Machine learning,
- Multivariate modeling and supervised learning,
- Dimension reduction techniques and unsupervised learning,
- Deep learning,
- Model assessment and sensitivity analysis, and
- Model interpretation (particularly for black box models).

Workflow and Reproducibility

Modern data science has at its core the creation of workflows—pipelines of processes that combine simpler tools to solve larger tasks. Documenting, incrementally improving, sharing, and generalizing such workflows are an important part of data science practice owing to the team nature of data science and broader significance of scientific reproducibility and replicability. Documenting and sharing workflows enable others to understand how data have been used and refined and what steps were taken in an analysis process. This can increase the confidence in results and improve trust in the process as well as enable reuse of analyses or results in a meaningful way.

Students need to be exposed to the concept of workflows and gain experience constructing them. Understanding the end-to-end structure of a workflow and being able to describe and document the workflow is important. Students need to learn about software systems that enable building workflows (e.g., R and Python) and how to document what they do (e.g., R Markdown and Jupyter Notebook). Studying end-to-end properties of workflows and then incrementally improving them in an evidence-based fashion is important. Students need to learn about such practices and learn how to execute such practices autonomously.

Providing experiential learning at multiple time points is important as students learn workflow processes and practice implementing and documenting steps within a workflow. Students need practice developing a unified approach to analysis and integration of multiple methods applied to data sets in an iterative manner. Project management could be integrated into capstone experiences. Longer-term projects involving interim reports and evaluation are critical.

Key workflow and reproducibility concepts/skills that would be important for all students in their data science programs and critical for their success in the workforce are the following:

- Workflows and workflow systems,
- Documentation and code standards,
- Source code (version) control systems,
- Reproducible analysis, and
- Collaboration.

Communication and Teamwork

One major distinguishing attribute of the work of data scientists centers on their capacity to frame research questions well and then communicate the findings in writing, in graphical form, and in conversation. In many cases, this involves coordinating among multidisciplinary actors, translating the interests of various parties, and then synthesizing the findings for nonexpert audiences. This requires competency in statistics, computer science, mathematics, coding, and domain-specific interests. Graduates also need to write clearly, speak articulately, construct effective visual displays and compelling written summaries, and communicate complex data science results in basic terms to various stakeholders.

The development of responsible oral and written communication skills is also essential for productive collaboration in the classroom and in the workplace. The ability to work well in multidisciplinary teams is a key component of data science education that is highly valued by industry, as teams of individuals with particular skill sets each play a critical role in producing data products. Multidisciplinary collaboration provides students with the opportunity to use creative problem solving and to refine leadership skills, both of which are essential for future project organization and management experiences in the workplace. Multidisciplinary teamwork also emphasizes inclusion and encourages diversity of thought in approaching data science problems.

Key communication and teamwork concepts/skills that would be important for all students in their data science programs and critical for their success in the workforce are the following:

- Ability to understand client needs,
- Clear and comprehensive reporting,
- Conflict resolution skills,
- Well-structured technical writing without jargon, and
- Effective presentation skills.

Domain-Specific Considerations

Effective application of data science to a domain requires knowledge of that domain. Grounding data science instruction in substantive contextual examples (which will require the development of judgment and background in those areas) will help ensure that data scientists develop the capacity to pose and answer questions with data. Reinforcing skills and capacities developed in data science courses in the context of a specific domain will help students see the entire data science process. This might include completion of a track in a domain area, specialized connector courses that link data science concepts directly to students' fields of interest to build data science skills in context, a minor in a domain area, or a co-major or double major in an application area. Hopefully, such an interconnected appreciation for a domain and for data science methods will generalize to other domains and applications.

Students who have completed the course Data 8: Foundations of Data Science at the University of California, Berkeley, for example, have an opportunity to enroll in specialized connector courses that are offered by a variety of academic departments. Examples of these connector courses include Data Science for Smart Cities, Making Sense of Cultural Data, Data Science and the Mind, and Data Science, Demography, and Immigration.[5] Students at the University of Illinois, Urbana-Champaign, also

---

[5] The website for the connector curriculum is https://data.berkeley.edu/education/connectors, accessed February 20, 2018.

have an opportunity to integrate their domain knowledge with data science concepts in the CS+X degree program. In this bachelor's degree program, students dedicate half of their coursework to the study of computer science and the other half to a specific discipline—current selections include mathematics, statistics, anthropology, astronomy, chemistry, linguistics, music, philosophy, geoscience, crop science, and advertising.[6] Such approaches reinforce the integrative nature of data science, offering a comprehensive educational experience that better prepares students for the future workforce. The committee anticipates that the demand for interdisciplinary experiences will increase as the field of data science continues to evolve. Additional interdisciplinary pairings—such as English and data science to prepare future data journalists—are likely to emerge.

Ethical Problem Solving

As powerful analytical tools are growing to meet new possibilities of collecting data, students need to be aware of ethical challenges that can emerge. With this proliferation of data and advancement of innovation, data science practitioners may often be confronted with decisions about whether they should take certain actions just because they have the ability and tools to do so. The explosion of data potentially raises the possibilities of new intrusions and interventions in people's lives and other previously "safe" and protected places. The misuse of data can pierce basic human dignities or thwart human agency and autonomy. Students working with data need to know the ways in which their findings might compromise people's dignity and their identities. Most disturbing of all, data can be misused in ways that are socially unjust. Students also need to be aware of legal requirements aimed at protecting individuals' privacy such as the European Union General Data Privacy Regulation, which aims to increase the rights of data subjects and provides penalties for individuals or organizations that violate them.[7]

Ethical considerations, in other words, lie at the heart of data science. Unique ethical considerations arise in each step of and throughout the data science life cycle (i.e., when posing a question; collecting, cleaning, and storing data; developing tools and algorithms; performing exploratory analysis and visualization; making inferences and predictions; making decisions; and communicating results). Stand-alone courses on ethics could help students learn what intelligent systems and the tools of data

---

[6] The website for the CS+X program is https://cs.illinois.edu/academics/undergraduate/degree-program-options/cs-x-degree-programs, accessed February 20, 2018.

[7] The website for the European Union General Data Protection Regulation is https://www.eugdpr.org/, accessed March 29, 2018.

science can and cannot do. It is important to emphasize to students that this is not simply a case of "do it like it is done today"—it is a case where ongoing improvement and elevation of standards are needed. Beyond the stand-alone ethics course, students stand to develop a deeper understanding of the role that ethics plays throughout the study and practice of the data science life cycle if ethical principles are incorporated into most of the courses in the data science curriculum.

Case studies may be an especially effective approach. For example, case studies could be used to show how vulnerable people can be exploited by means of their medical or behavioral data being shared. Through these case studies, students could begin to develop a sense of awareness of the potential impacts on inequality of the misuse of data. In addition to learning about standards for responsible behavior through such case studies, students would also benefit from instruction in developing specific skills to navigate the challenging ethical problems with which data scientists struggle.

Key aspects of ethics needed for all data scientists (and for that matter, all educated citizens) include the following:

- Ethical precepts for data science and codes of conduct,
- Privacy and confidentiality,
- Responsible conduct of research,
- Ability to identify "junk" science, and
- Ability to detect algorithmic bias.

Recommendation 2.4: Ethics is a topic that, given the nature of data science, students should learn and practice throughout their education. Academic institutions should ensure that ethics is woven into the data science curriculum from the beginning and throughout.

A CODE OF ETHICS FOR DATA SCIENCE

Other disciplines have benefited from publishing specific ethical guidelines by which their members agree to conduct themselves. Practitioners in the fields of medicine and engineering have long traditions of similar ethical guidelines. The *AMA Code of Medical Ethics* of the American Medical Association includes guidance on interactions between medical professionals and their patients; use of medical treatments, including those that rely on new technologies; and "self-regulation" within the workplace (AMA, 2016). The "IEEE Code of Ethics" of the Institute of Electrical and Electronic Engineers encourages engineering professionals to prioritize the safety of the public, avoid or disclose conflicts of interest, present evidence-based claims, and maintain appropriate technical qualifications, for example (IEEE, 2017).

Considerable work in the study of ethical decision making for scientists has been undertaken by the Association of Computing Machinery (ACM) and the ASA. The *2018 ACM Code of Ethics and Professional Conduct: Draft 3* (an update to the ACM's 1992 code of ethics) includes principles for moral conduct in addition to leadership guidelines for computing professionals acting in the interest of the public good (ACM, 2018). The ASA's *Ethical Guidelines for Statistical Practice* presents guidelines pertaining to integrity and accountability in statistical work. It also details the various ethical responsibilities that statisticians have toward their research subjects, clients, employers, and colleagues (ASA, 2016).

These rules of conduct uphold specific ethical standards for professionals whose activities and practice can significantly impact the health and well-being of people, society, and their profession. As an emerging discipline, data science could benefit from having its own ethical standards of conduct. There are many areas specific to data science that could be addressed, including the responsibility to protect privacy of personal data, the responsibility to not misrepresent the data for personal gain, the responsibility to ensure fairness in the use of machine learning algorithms and choice of training data, and the responsibility to ensure that results produced by the analyst are reproducible.

Given the sensitive nature of certain types of data and the significant ethical implications of working with such data, efforts to establish a code of ethics for data scientists are under way throughout the field.[8] Data science ethics might be codified in an "oath" similar to the Hippocratic Oath taken by physicians as a way to crystallize what is being asked of them. Although the specific content and form of an oath may be controversial, it can also underline the importance of the commitment being made. A draft version of such an oath was presented in the interim report from this committee, and a revised version appears in Appendix D of this report. The potential consequences of the ethical implications of data science cannot be overstated.

---

[8] To read about other work in the development of data science codes of ethics, see, for example, https://datapractices.org/community-principles-on-ethical-data-sharing/, http://datafordemocracy.org/projects/ethics.html, http://www.datascienceassn.org/code-of-conduct.html, http://www.rosebt.com/blog/open-for-comment-proposed-data-science-code-of-professional-conduct, https://dssg.uchicago.edu/2015/09/18/an-ethical-checklist-for-data-science/, http://thedataist.com/a-proposal-for-data-science-ethics/, https://www.accenture.com/t20160629T012639Z__w__/us-en/_acnmedia/PDF-24/Accenture-Universal-Principles-Data-Ethics.pdf, accessed January 31, 2018.

Recommendation 2.5: The data science community should adopt a code of ethics; such a code should be affirmed by members of professional societies, included in professional development programs and curricula, and conveyed through educational programs. The code should be reevaluated often in light of new developments.

REFERENCES

ACM (Association for Computing Machinery). 2018. *2018 ACM Code of Ethics and Professional Conduct: Draft 3*. https://ethics.acm.org/2018-code-draft-3. Accessed February 6, 2018.

AMA (American Medical Association). 2016. *AMA Code of Medical Ethics*. https://www.ama-assn.org/delivering-care/ama-code-medical-ethics. Accessed February 12, 2018.

ASA (American Statistical Association). 2014. *Curriculum Guidelines for Undergraduate Programs in Statistical Science*. http://www.amstat.org/asa/files/pdfs/EDU-guidelines2014-11-15.pdf.

ASA. 2015. *ASA Statement of the Role of Statistics in Data Science*. http://ww2.amstat.org/misc/DataScienceStatement.pdf.

ASA. 2016. *Ethical Guidelines for Statistical Practice*. http://www.amstat.org/asa/files/pdfs/EthicalGuidelines.pdf.

BHEF and PwC (Business-Higher Education Forum and PricewaterhouseCoopers). 2017. *Investing in America's Data Science and Analytics Talent: The Case for Action*. http://www.bhef.com/sites/default/files/bhef_2017_investing_in_dsa.pdf.

Butler, D. 2013. When Google got flu wrong. *Nature* 494:155-156.

Chen, H., R.H.L. Chiang, and V.C. Storey. 2012. Business intelligence and analytics: From big data to big impact. *MIS Quarterly* 36(4):1165-1188.

Chin, J., and L. Lin. 2017. China's all-seeing surveillance state is reading its citizens' faces. *Wall Street Journal*, June 26. https://www.wsj.com/articles/the-all-seeing-surveillance-state-feared-in-the-west-is-a-reality-in-china-1498493020.

Codella, N.C.F., Q.B. Nguyen, S. Pankanti, D. Gutman, B. Helba, A. Halpern, and J.R. Smith. 2017. Deep learning ensembles for melanoma recognition in dermoscopy images. *IBM Journal of Research and Development* 61(4):5.1-5.15.

Columbus, L. 2017. IBM predicts demand for data scientists will soar 28% by 2020. *Forbes*, May 13.

CRA (Computing Research Association). 2016. *Computing Research and the Emerging Field of Data Science*. https://cra.org/wp-content/uploads/2016/10/Computing-Research-and-the-Emerging-Field-of-Data-Science.pdf.

Danyllo, W.A., V.B. Alisson, N.D. Alexandre, L.M.J. Moacir, B.P. Jansepetrus, and R.F. Oliveira. 2013. "Identifying Relevant Users and Groups in the Context of Credit Analysis Based on Data from Twitter." Paper presented at the 2013 IEEE Third International Conference on Cloud and Green Computing, September/October, Karlsruhe, Germany.

De Veaux, R., M. Agarwal, M. Averett, B.S. Baumer, A. Bray, T.C. Bressoud, L. Bryant, et al. 2017. Curriculum guidelines for undergraduate programs in data science. *Annual Review of Statistics and Its Applications* 4:15-30.

Donoho, D. 2017. 50 years of data science. *Journal of Computational and Graphical Statistics* 26(4):745-766.

Ernst and Young. 2017. "Data and Advanced Analytics: High Stakes, High Rewards." *Forbes Insights*, February. https://www.forbes.com/forbesinsights/ey_data_analytics_2017/. Accessed February 13, 2018.

Hardin, J.S., and N.J. Horton. 2017. Ensuring that mathematics is relevant in a world of data science. *Notices of the AMS* 64(9):986-990. https://www.ams.org/publications/journals/notices/201709/rnoti-p986.pdf.

Hvistendahl, M. 2016. Can "predictive policing" prevent crime before it happens? *Science*, October 5. http://www.sciencemag.org/news/2016/09/can-predictive-policing-prevent-crime-it-happens.

IEEE (Institute of Electrical and Electronics Engineers). 2017. "IEEE Code of Ethics." https://www.ieee.org/about/corporate/governance/p7-8.html. Accessed February 12, 2018.

Jordan, M. 2013. On statistics, computation and scalability. *Bernoulli* 19(4):1378-1390.

Kitchin, R. 2014. The real-time city? Big data and smart urbanism. *GeoJournal* 79:1-14.

Markow, S., S. Braganza, B. Taska, S. Miller, and D. Hughes. 2017. *The Quant Crunch: How the Demands for Data Science Skills Is Disrupting the Job Market*. https://www-01.ibm.com/common/ssi/cgi-bin/ssialias?htmlfid=IML14576USEN&. Accessed June 21, 2017.

NRC (National Research Council). 2013. *Frontiers in Massive Data Analysis*. Washington, D.C.: The National Academies Press.

NRC. 2014. *Training Students to Extract Value from Big Data: Summary of a Workshop*. Washington, D.C.: The National Academies Press.

Pratt, M.K. 2016. Big data's big role in humanitarian aid. *Computer World*, February 8. http://www.computerworld.com/article/3027117/big-data/big-datas-big-role-in-humanitarian-aid.html. Accessed June 21, 2017.

UC Santa Cruz (University of California, Santa Cruz). 2018. "Program Learning Outcomes: Programs, Curriculum Alignment, and Assessment Plans. Jack Baskin School of Engineering." https://www.soe.ucsc.edu/departments/computer-science/program-learning-outcomes. Accessed January 18, 2018.

Wing, J.M. 2006. Computational thinking. *Communications of the ACM* 49(3):33-35.

3

Data Science Education

Data science educational opportunities for students are rapidly growing. Universities, colleges, community colleges, and other organizations are starting to offer a range of programs for students with different interests and backgrounds. This chapter provides an overview of the current landscape of classes and programs and identifies some of the key challenges facing those who aim to develop a data science program.

UNDERGRADUATE MODALITIES

Undergraduate data science education is currently offered in many forms, and this variability is expected to continue in the near future. Common modalities include the following:

- Introductory exposure to data science, through a single inspirational course that could satisfy a general education requirement;
- Major in data science, including advanced skills, as the primary field of study;
- Minor or track in data science, where intermediate skills are connected to the major field of study;
- Two-year degrees and certificates;
- Other certificates;
- Massive open online courses (MOOCs), which can engage large numbers of students at a variety of levels; and

- Summer programs and boot camps, which can serve to supplement academic or on-the-job training.

As academic institutions add courses and programming around undergraduate data science, they will need to decide what modalities are institutionally appropriate, considering many factors such as student demand, faculty and institutional strengths and resources, and curricular fit. These choices may also be influenced by the existence of graduate programs in data science at the institution. Each of these modalities—with its strengths, limitations, and possible areas for improvement—is discussed in more detail in the following sections.

Introductory Exposure to Data Science

Several 4-year and, increasingly, 2-year academic institutions have some type of introductory data science course or sequence to educate students, and many more institutions will likely develop these courses in years to come, especially as degree programs are developed for students. These stand-alone courses and sequences provide interested students with an introduction to data science, attract students to data science majors or minors (if applicable), and broadly prepare students for an increasingly data-driven world.

Some institutions are implementing a general education offering or requirement in data science, which may be taught by multiple departments and offered via semester-long courses, modules, virtual sessions, self-guided instruction, or MOOCs. This flexibility ensures that data science may be integrated into other programs of study. Other institutions are providing introductory courses to meet the demands of interested students and to better prepare them to live and work in a world in which it is essential to know how to engage with data critically and carefully. Transferable data science skills include descriptive statistics, visualization, sampling, programming, managing databases, inference, and business analytics. Students from various disciplines seem eager to enroll in data science courses in part because the content may add value to their degrees.

The following are a few examples of institutions that provide an introductory data science experience:

- The University of California, Berkeley, a public research university, offers an introductory data science course, Data 8: Foundations of Data Science.[1] This course is open to all students, regardless of

---

[1] The website for Data 8: Foundations of Data Science is http://data8.org, accessed January 25, 2018.

educational backgrounds or majors, and there are no prerequisites to enroll beyond those required for entry to the university. The course is cross-listed in the Department of Computer Science, the Department of Statistics, and the School of Information and is taught by an interdisciplinary team of faculty. Reflecting the level of student interest, it has grown from a pilot of less than 100 students in fall 2015 to over 1,100 in spring 2018, drawing from over 70 different majors (UC Berkeley, 2018). Each term, about a dozen wide-ranging "connector courses" are offered concurrently with the introductory course to connect it with areas of student interest, such as legal studies, cognitive neuroscience, geography, history, civil engineering, immunology, demography, psychology, business, and others.[2] To further increase accessibility, the entire course experience is through Jupyter notebooks in the cloud, so students need only a web browser to participate fully.

- Amherst College, a private liberal arts college in Massachusetts, also offers an introductory data science course through its Mathematics and Statistics Department, STAT-231: Data Science.[3] The course "provides a practical foundation for students to think with data by participating in the entire data analysis cycle. Students will generate statistical questions and then address them through data acquisition, cleaning, transforming, modeling, and interpretation . . . [and will use and apply] tools for data management and wrangling that are common in data science . . . to real-world applications" (Amherst College, 2017). The course has a prerequisite of some background in statistics and computer science.
- Carnegie Mellon University, a private research university in Pittsburgh, Pennsylvania, offers Reasoning with Data, a first course in statistics and data science focusing on concepts, interpretation, and communication. It is part of the required general education curriculum for all students in the Dietrich College of Humanities and Social Sciences. Students use an interactive platform that allows for analysis without using a specific programming language. The coursework includes several student-driven data analysis projects.[4]
- The University of Washington, a public research university in Seattle, offers Introduction to Data Science, which is cross-listed in

---

[2] The website for the UC Berkeley connector courses is https://data.berkeley.edu/education/connectors, accessed February 12, 2018.

[3] The website for STAT-231 is https://www.amherst.edu/academiclife/departments/courses/1718F/STAT/STAT-231-1718F, accessed January 25, 2018.

[4] To view a sample syllabus from Reasoning with Data, see http://www.stat.cmu.edu/~rnugent/PUBLIC/teaching/200syllabus.pdf.

the Department of Statistics, the Department of Computer Science and Engineering, and the Information School.[5] It uses the interactive textbook *Computational and Inferential Thinking: The Foundations of Data Science* (Adhikari and DeNero, 2018) and the Jupyter-based notebook environment of the University of California, Berkeley, course Data 8: Foundations of Data Science. Introduction to Data Science requires only precalculus as a prerequisite, and it addresses data collection and management, summary and visualization of data, basic statistical inference, and machine learning.

- Winona State University, a public university in Minnesota, offers an introductory data science course, DSCI 210: Data Science, that allows students to explore methods and techniques commonly used by data scientists. Participants have an opportunity to learn about data management, preparation, analysis, visualization, and modeling as well as to complete a data science project. Students are required to complete an introductory computing course as a prerequisite to enrolling in this class. DSCI 210 serves as an introduction to data science for nonmajors, but it also counts toward the requirements for Winona State's B.S. in data science.[6]

- The computer science department at St. Olaf College, a private liberal arts college in Northfield, Minnesota, hosts the introductory course CSCI 125: Computer Science for Scientists and Mathematicians.[7] In this course, students discuss how to handle, visualize, find patterns in, and communicate about data. Students have the opportunity to learn how to use common data science tools (e.g., Python and R) while working collaboratively on real-world problems. To enroll in the course, students need to have previous coursework in calculus.

- Montgomery College, a public 2-year college with campuses across Maryland, offers DATA 101: Introduction to Data Science.[8] Students are required to have taken one of four approved statistics courses as a prerequisite to enroll in this course. Throughout the course, students explore methods to collect, organize, manage, examine, prepare, analyze, and visualize data. This introductory course can

---

[5] To view a sample syllabus from Introduction to Data Science, see https://wstuetzle.github.io/IDS-syllabus-11-14-2017.html, accessed February 20, 2018.

[6] The website for DSCI 210 and the B.S. in data science at Winona State is https://catalog.winona.edu/preview_program.php?catoid=10&poid=2474&returnto=958, accessed February 1, 2018.

[7] The website for St. Olaf College's computer science department is http://catalog.stolaf.edu/academic-programs/computer-science/, accessed February 1, 2018.

[8] The website for DATA 101 is http://catalog.montgomerycollege.edu/preview_course_nopop.php?catoid=8&coid=11413, accessed February 1, 2018.

also be used to satisfy one of the requirements for Montgomery College's Data Science Certificate, which adds courses on writing and communicating about data and using statistical methods, as well as a capstone experience.

Introductory data science experiences can also attract students to other data science programs offered by the institution, such as majors, minors, tracks, and certificates. These students include members of other disciplines, such as the humanities, social sciences, and the arts, as well as members of populations underrepresented in science, technology, engineering, and mathematics fields. Introductory data science courses offer a low-stakes opportunity (i.e., without barriers related to previous training or expertise) to investigate the field of data science and to be exposed to the skills and experiences that would be applicable to a wide range of future careers. Introductory courses serve as a possible gateway to degrees or specializations in data science, as they motivate students to see the valuable role that data science expertise will play in the future workforce.

Developing, implementing, and delivering an introductory data science course is not without its challenges, however. For example, while the prospect of offering introductory courses without prerequisites is most attractive for the purposes of broadening participation in data science, unrestricted enrollment must be reconciled with classroom capacity and instructor availability. Content is also likely to vary by course, owing to instructor expertise and interest and by student knowledge; this variability may create issues for students in terms of consistent preparation for future courses. These issues also confront developers of upper-level elective courses or of courses relating to a data science major, as it could be difficult to prepare content and develop learning outcomes for a population of students with such varied levels of knowledge and experience.

Introductory data science courses will continue to evolve and improve over time with innovative curriculum development that emphasizes real-world experiences and connections, ongoing faculty development opportunities, and cross-disciplinary collaboration among a wider spectrum of disciplines.

Major in Data Science

Data science majors are emerging across academic institutions and will continue to do so in years to come. Similar to introductory data science experiences, there is significant variation in program structure, goals, and content in these majors as well. Some data science majors are emerging as independent programs that interface with specific domain areas, while others are emerging as specializations within a given domain area.

The most common features of current data science majors include required courses in mathematics, statistics, and computer science. Within mathematics departments, requirements often include the following:

- Mathematics courses on linear algebra, calculus, and discrete structures;
- Statistics courses on introductory statistics, probability, and various kinds of applied statistics; and
- Computer science courses on database systems, programming, data structures, algorithms, and machine learning.

Some majors have courses listed as "data science," including cross-listed courses with statistics and computer science, while other data science majors draw entirely upon courses from connected departments. Common topics taught under the "data science" listing include advanced data analytics, big data, data mining, simulation modeling, and computational thinking. Many data science majors include required or elective courses from outside the core departments—commonly economics, business, psychology, biology, and geography or geosciences. Many data science majors also require a hands-on practicum or capstone course to help reinforce skills.

Currently, many 4-year majors fall into one of three categories: (1) data science majors housed within a college or school of business (i.e., programs in business analytics, which usually involve more marketing and finance classes and fewer computational and mathematics courses); (2) data science/analytics majors housed in a mathematics or statistics department (i.e., above-average mathematics or statistics requirements with fewer "core" computational courses); and (3) data science programs housed in a computer science department as either a stand-alone major or as a concentration to information technology (i.e., more computational courses but potentially fewer "core" mathematics courses). Variations in courses offered and required within similarly labeled majors at different institutions are notable. A few 4-year undergraduate data science majors are hybrids of these three models, being administered jointly by multiple departments. The following list includes a variety of approaches to data science majors:

- The University of Michigan, a public research university in Ann Arbor, launched a major in data science in fall 2015. This major is a joint program offered by two departments in two different colleges: computer science in the College of Engineering and statistics in the College of Literature, Sciences, and the Arts (LSA).[9] The major requirements consist of a core of five courses: discrete mathematics, programming, data structures, probability and statistics, and applied regression. In addition to satisfying the core requirements, students select at least one course from each of three areas: machine learning, data management, and data science applications. Engineering students majoring in data science also take a computer professionalism course. Both LSA and engineering data science majors participate in a capstone experience, typically during their senior year.

- Smith College, a private liberal arts institution in Northampton, Massachusetts, began offering a major in statistical and data sciences in fall 2017.[10] The program is not hosted by one campus department; instead, it draws on faculty and disciplines from across the college. The major requires 10 courses, including courses in statistics, computer science, data science, communication, and a domain area.
- Virginia Tech, a public research and land-grant university in Blacksburg, offers a major in computational modeling and data analytics in its College of Science.[11] The major includes courses in mathematics, statistics, computer science, and data science, as well as a capstone experience.
- The University of California, San Diego, a public research university, launched a B.S. in data science in fall 2017 through its departments of cognitive science, computer science and engineering, and mathematics.[12] The major consists of a technical lower division comprising mathematics, computer science and engineering, natural sciences, and five specific data science courses. The upper division has core mathematics, computer science, and data science courses; elective computer science and mathematics courses; and a senior project.[13] On completion of the major, students need to be "versed in predictive modeling, data analysis and computational

---

[9] The websites for the joint program at the University of Michigan are https://www.eecs.umich.edu/eecs/undergraduate/data-science/ and https://lsa.umich.edu/stats/undergraduate-students/undergraduate-programs/majordatascience.html, accessed February 12, 2018.

[10] The website for the Smith College major in statistical and data sciences is https://www.smith.edu/statistics/about.php, accessed January 25, 2018.

[11] The website for the Virginia Tech major in computational modeling and data analytics is https://www.ais.science.vt.edu/programs/cmda.html, accessed January 25, 2018.

[12] The website for the University of California, San Diego, data science major is http://dsc.ucsd.edu, accessed January 25, 2018.

[13] The website describing the course requirements for the B.S. in data science is http://dsc.ucsd.edu/node/7, accessed February 12, 2018.

techniques ... [and have developed] undergraduate-level expertise in a specific subject area outside of data science" (UC San Diego, 2017).

- At the University of Rochester, a private university in New York state, the Goergen Institute for Data Science offers a major in data science in the form of either a B.A. or a B.S.[14] While this major includes studies in computer science and statistics, as well as a capstone experience, it also allows students to take upper-level coursework in a focused domain area such as business, environmental science, biology, or political science.
- The Massachusetts Institute of Technology, a private research university in Cambridge, introduced a B.S. in computer science, economics, and data science. It seeks to equip students with a foundational knowledge of economic analysis, computing, optimization, and data science, as well as hands-on experience with empirical analysis of economic data to identify, analyze, and solve real-world challenges in real and virtual settings.[15] Required courses are drawn from lists of preexisting courses in mathematics, electrical engineering and computer science, and economics covering linear algebra, discrete mathematics, probability and statistics, computation and algorithms, data science, intermediate economics, elective subjects from data science and economics theory, and communications practice.
- The University of California, Irvine, a public research university, houses a B.S. in data science in its Statistics Department, within the Bren School of Information and Computer Sciences. It consists of 16 lower division quarter courses, including nine computer science courses, four mathematics courses, and three statistics courses. Two of the three statistics courses are specifically related to data science. The upper division requirements include seven statistics courses, three computer science (including machine learning) courses, one writing course, one visualization course, three approved elective courses from mathematics/computer science/information, and a two-quarter capstone course.[16]
- The New York University School of Professional Studies offers a B.S. in applied data analytics and visualization. Its core curriculum

---

[14] The website for the Goergen Institute for Data Science is http://www.sas.rochester.edu/dsc/undergraduate/index.html, accessed January 25, 2018.

[15] The website for the MIT B.S. in computer science, economics, and data science is https://www.eecs.mit.edu/academics-admissions/undergraduate-programs/6-14-computer-science-economics-and-data-science, accessed February 20, 2018.

[16] The website for the University of California, Irvine, data science degree is http://www.ics.uci.edu/ugrad/degrees/degree_datascience.php, accessed February 20, 2018.

consists of general education courses (e.g., writing, critical thinking, quantitative reasoning, scientific issues, and history, art, and culture) and liberal arts electives, with the following major requirements: mathematics (i.e., precalculus, calculus, and linear algebra), statistics, computer science (i.e., fundamentals, database design, networking, and systems analysis), eight specific data science and visualization courses, three electives in applied analytics and visualization, and a graduation project.[17]

Challenges in forming a major program include constructing gateways from established lower division courses to the advanced courses in the major, forming a set of upper division courses that covers the essentials of data science without essentially forming a double or triple major, incorporating a domain of application, addressing ethical issues and social implications, and considering a hands-on practicum or capstone integrative experience. Preexisting courses from computer science, statistics, applied mathematics, operations, and information management typically cover essential material but do so in the context of those majors. Forming new courses that integrate segments of several such courses may streamline the data science major but may overlap to some extent with existing course offerings. Thus, it is particularly important that the unique data science character be brought out when forming such amalgams.

Several institutions have developed courses that address these challenges in various ways. For example, Carnegie Mellon University's statistics program builds on its introductory Reasoning with Data course with Methods for Statistics and Data Science. It focuses on regression and nonparametric methods, while requiring the use of R Markdown and GitHub in structured coding templates. Students design data analysis projects and write reports. Carnegie Mellon University's computer science program offers an independent course, Practical Data Science,[18] focused on data collection and processing, data visualization and presentation, statistical model building using machine learning, and big data techniques for scaling these methods.

Another example is from the University of California, Berkeley, which introduced the junior-level Principles and Techniques of Data Science.[19]

---

[17] The curriculum for the New York University B.S. in applied data analytics and visualization can be found at http://www.sps.nyu.edu/academics/departments/mcghee/undergraduate/bachelors/bs-applied-data-analytics-and-visualization/core-major-curriculum.html, accessed February 12, 2018.

[18] The website for Practical Data Science is http://www.datasciencecourse.org/, accessed February 20, 2018.

[19] The website for Principles and Techniques of Data Science is http://www.ds100.org/, accessed February 20, 2018.

This course seeks to provide a second integrated experience of computational and inferential thinking that builds on the introductory foundational courses and adds mathematics background. The course is oriented around the data science life cycle. Rather than relying on the tool- and environment-agnostic pedagogy at the introductory level, this more advanced course provides direct exposure to current data science technology. It also seeks to provide background necessary for students to take advanced computer science and statistics courses that are particularly relevant to data science, without all of the conventional prerequisites of those majors. Topics include languages for transforming, querying, and analyzing data; algorithms for machine learning methods including regression, classification, and clustering; principles behind creating informative data visualizations; statistical concepts of measurement error and prediction; and techniques for scalable data processing.

Minor or Track in Data Science

For those undergraduate students who are not interested in or do not have the time to complete a 4-year major in data science, a minor or track in data science offers a useful alternative for introductory skill development. This may be a good approach for students in other domain-specific programs who want to gain data science expertise while remaining firmly in their chosen domain. In such programs, students are exposed to data science concepts, tools, and techniques that are related to their chosen discipline. A number of academic institutions offer minors or tracks in data science; the following is only a small selection of these programs:

- The Georgia Institute of Technology, a public research university in Atlanta, offers a minor in computational data analysis through its College of Computing.[20] Required courses include Introduction to Computing for Data Analysis, Data and Visual Analytics, an approved introductory probability and statistics course, an approved computational methods course, and an approved computational data analysis elective.
- The University of Massachusetts Amherst, a public research university, offers a B.A. track in informatics in data science through its College of Information and Computer Sciences.[21] Required courses include Mathematical Foundations of Informatics, an approved

---

[20] The website for the Georgia Tech minor in computational data analysis is https://www.cc.gatech.edu/content/minor-computational-data-analysis, accessed January 25, 2018.

[21] The website for the University of Massachusetts Amherst track in informatics is https://www.cics.umass.edu/informatics, accessed January 25, 2018.

introductory statistics course, Problem Solving with Computers, Networked World, Human Computer Interaction, Web Programming, Databases, Informatics, Using Data Structures, Data Science, and approved electives in statistics, data science, social science, and/or business.
- The University of Washington offers a Data Science Option[22] within several undergraduate programs, where each department may specialize its option within a common template. An option includes one course each in programming, machine learning, and societal implications of data science, as well as a course in two of three areas: data management, data visualization and communication, and advanced statistics and probability. Optionally, it may require Introduction to Data Science.
- Westminster College, a private liberal arts institution in Salt Lake City, Utah, offers a data science minor out of its Data Science Department.[23] Required courses include Linear Algebra, Introduction to Statistics, either Introduction to Computer Science or Scientific Computing, Explorations in Data Science, two approved electives (both from mathematics, computer science, or statistics), and a capstone.
- Miami University of Ohio, a public research university in Hamilton, offers a business analytics minor hosted by its business department.[24] Required courses include Business Statistics, Applied Regression and Analysis in Business, IT and the Intelligent Enterprise, Database Systems and Data Warehousing, Business Intelligence and Data Visualization, and two approved electives from data science, social science, and/or statistics.
- The Rose-Hulman Institute of Technology, a private college in Terre Haute, Indiana, offers a multidisciplinary minor in data science.[25] Required courses include an approved introductory statistics course, Introduction to Software Development, Object-Oriented Software Development, and four electives from an approved list of statistics, computer science, and data science courses.

---

[22] The website for the Data Science Option is https://www.cs.washington.edu/academics/ugrad/courses/data-science, accessed February 20, 2018.

[23] The website for the Westminster College data science minor is https://catalog.westminstercollege.edu/undergraduate2016-2017/data-science-data/, accessed January 25, 2018.

[24] The website for the Miami University of Ohio business analytics minor is https://miamioh.edu/fsb/academics/isa/academics/minors/business-analytics/index.html, accessed January 25, 2018.

[25] The website for the Rose-Hulman multidisciplinary data science minor is https://www.rose-hulman.edu/academics/course-catalog/current/special-programs.html#multi-disc-minor-data-science, accessed January 25, 2018.

- Louisiana Tech University, a public research university in Ruston, offers a B.S. track in computer science, cloud computing, and big data through its College of Engineering and Science.[26] Required courses include Calculus, Discrete Mathematics, Statistical Methods, Science of Computing, Systems Programming, Digital Design, Operating Systems, Theory of Computing, Data Structures, Advanced Data Structures and Algorithms, Programming Languages, Computer Architecture, Database Management Systems, Software Design and Engineering, Distributed and Cloud Computing, Computer Networks or Artificial Intelligence, Data Mining and Knowledge Discovery, and Advanced Data Mining, Fusion, and Application, plus approved electives in computer science and a capstone course.

Minors or tracks in data science give students a wider knowledge base, thus making them more marketable candidates for the future workforce—not only do these students have expertise in a particular domain area, but they also have valuable data science skills and experience that can be applicable to a variety of emerging career paths. Although the minor or track in data science is an attractive option for certain populations of students, such programs may be limited by time and space in their ability to help students develop data science foundations and skills in sufficient depth.

Two-Year Degrees and Certificates

Many 2-year institutions are starting data science programs using a wide variety of educational approaches, ranging from a few courses leading to a certificate to a full associate's degree requiring 2 years of study. The blend among statistics, computer science, and applications is highly variable depending on the program. These programs will often simultaneously serve to (1) be an entry point to inspire and attract a wide variety of student populations to data science; (2) permit existing members of the workforce to retrain or obtain specific new skill sets to complement their education and experience; (3) create mechanisms by which students can certify specific or general skill sets with certificates or associate's degrees; (4) build foundational, translational, ethical, and professional skills to support matriculation into 4-year data science programs; and (5) provide opportunities for advanced high school students to begin data science training early. The majority of these purposes support undergraduate

---

[26] The website for the Louisiana Tech track in computer science, cloud computing, and big data is http://catalog.latech.edu/preview_program.php?catoid=3&poid=913, accessed January 25, 2018.

education objectives, while also targeting the specific needs of industry. For example, a data management program may focus largely on data systems, collection, and curation; a data analytics program may focus more on algorithms, statistics, and machine learning; and a business analytics program may focus on business and supply chain issues. Currently, most programs offer courses on data visualization, but relatively few programs offer courses on writing about data (communication in data science). A few of these programs are online only, while most have some online component. More than half of the 2-year institutions studied by the committee have their own unit of data science offering the data science associate's degree or certificate, while programs that are hosted by other departments are typically in mathematics, business, or computer information systems. Some programs focus on training to acquire technical skills for industry and research through data science certificates (e.g., Montgomery College,[27] Normandale Community College,[28] Washtenaw Community College[29]). Others offer 2-year associate's degree programs in data science that prepare students to transfer into 4-year programs. Two examples of the latter are the following:

- The Community College of Allegheny County, with multiple Pennsylvania campuses, offers an associate's degree in data analytics technology[30] and prepares students to transfer to a 4-year institution with a data analytics bachelor's degree program or pursue employment in a variety of data science roles. Upon completion of their coursework, students need to be able to "identify and define data challenges in a variety of industries, collect and organize information from many sources, discover patterns and relationships in large data sets applying statistical tools, resolve business questions and make recommendations through data mining techniques, and communicate tactical and strategic objectives utilizing data visualization techniques" (CCAC, 2018).

---

[27] The website for the Montgomery College data science certificate is https://cms.montgomerycollege.edu/datascience/, accessed February 12, 2018.

[28] The website for the Normandale Community College certificate in data analysis is http://www.normandale.edu/continuing-education/data-analysis, accessed February 12, 2018.

[29] The website for the Washtenaw Community College certificate in applied data science is http://www.wccnet.edu/academics/programs/view/program/CTADS/, accessed February 12, 2018.

[30] The website for the Community College of Allegheny County data analytics technology program is https://www.ccac.edu/Data_Analytics_Technology.aspx, accessed February 1, 2018.

- Nashua Community College, located in New Hampshire, offers an associate's degree in foundations of data analytics.[31] This degree program is also designed for students who want to pursue a bachelor's degree in data analytics at a 4-year institution. Upon completion of this degree, students are expected to be able to "demonstrate technical proficiency and effective problem-solving ability in completing mathematical processes; identify the benefits of quality, timeliness, and continuous improvement in regards to software development; apply critical-thinking skills to identify, analyze, and solve problems; apply mathematical concepts to other disciplines including business, economics, social sciences, and the natural sciences; demonstrate the ability to follow a systematic progression of software development and refinement when designing and developing software for a project; participate effectively as a member of a team; articulate an understanding of the need for lifelong learning; demonstrate an understanding of diversity through interaction with project teammates; and develop software programs that reflect the application of up-to-date tools and techniques of the discipline" (Nashua Community College, 2018).

Two-year institutions throughout the United States have created certificate programs in data science. For instance, Maryland's Montgomery College (2018) has recently built a 2-year certificate program around courses in mathematics and statistics; fundamentals of coding, writing, and communication in data science; and a capstone project applying Cross-Industry Standard Process for Data Mining methodology to create "original yet reproducible analyses in a variety of formats for the general public and for members of the data science community." Outcomes from this program include a student's ability to do the following:

- Assess different analysis and data management techniques and justify the selection of a particular model or technique for a given task,
- Execute analysis of large and disparate data sets and construct models necessary for these analyses,
- Summarize findings of complex analyses in a concise way for a target audience using both graphics and statistical measures, and
- Demonstrate competency with programming languages and environments for data analysis (Montgomery College, 2018).

---

[31] The website for the Nashua Community College foundations of data analytics program is http://www.nashuacc.edu/academics/associate-degrees/stem-and-advanced-manufacturing/398-foundations-in-data-analytics, accessed February 1, 2018.

Although 2-year institutions confront many of the same obstacles in curricular and programmatic development as 4-year institutions, they also face many unique challenges. Administrators and faculty need to consider how best to develop content that will satisfy the needs of both students who may be interested in enrolling in only one course and those who plan to embark on a certificate program or an entire degree program. Furthermore, 2-year curricula are expected to engage students who seek an associate's degree and specific workforce training as well as students who need to develop a broader skill set and knowledge base in order to transfer to a 4-year bachelor's degree program. Two-year college faculty and administrators also have to be continuously aware of and responsive to the demands of the local workforce to be prepared to upskill both their transient students and established professionals as efficiently and as appropriately as possible.

Other Certificates

Other certificates in data science are proliferating across academic institutions and industry, including at the undergraduate level (e.g., the University of Georgia Applied Data Science Certificate Program;[32] the Temple University Certificate in Data Science: Computational Analytics;[33] the University of Missouri, St. Louis, Certificate in Data Science;[34] the Johnson County Community College Data Analytics Certificate;[35] and the Great Bay Community College Certificate in Practical Data Science[36]); the graduate level (e.g., Georgetown University's certificate in data science,[37]

---

[32] The website for the University of Georgia Applied Data Science Certificate Program is https://csci.franklin.uga.edu/certificate-applied-data-science, accessed February 6, 2018.

[33] The website for the Temple University Certificate in Data Science: Computational Analytics is http://bulletin.temple.edu/undergraduate/science-technology/computer-information-science/data-science-computational-analytics-certificate/, accessed February 6, 2018.

[34] The website for the University of Missouri, St. Louis, Certificate in Data Science is https://www.umsl.edu/mathcs/undergraduate-studies/certificatedatascience.html, accessed February 6, 2018.

[35] The website for the Johnson County Community College Data Analytics Certificate is http://catalog.jccc.edu/degreecertificates/computerinformationsystems/dataanalyticscert/, accessed February 6, 2018.

[36] The website for the Great Bay Community College Certificate in Practical Data Science is http://greatbay.edu/courses/certificate-programs/data-practical-data-science, accessed February 12, 2018.

[37] The website for the Georgetown University certificate in data science is https://scs.georgetown.edu/programs/375/certificate-in-data-science/?utm_source=Google&utm_medium=Search&utm_campaign=FY18_Search_Professional_Certificates&gclid=CjwKCAiAksvTBRBFEiwADSBZfI934rB_0ubuN0Zzmjr4xGtGmDLbei9zB5L8HN4MiLIzWEZN1BiKXBoC4ZIQAvD_BwE, accessed February 12, 2018.

the University of Michigan certificate in data science,[38] and others[39]); and the executive level (e.g., Columbia University[40] and Kellogg Executive Education at Northwestern University[41]). Online certificate options are also becoming common (e.g., edX MOOC[42] and Coursera[43]) and offer an accessible alternative to in-person learning. (MOOCs are discussed more in the next section.)

While certificate programs can be a valuable resource in upskilling current members of the workforce and supplementing other existing educational modalities, the program offerings are varied and often difficult to compare. Some certificates are granted after the successful completion of a single course, while others require the completion of a multicourse series. The lack of standardization can make it challenging for employers and prospective students to assess what the certificates represent and how they may fit within their data science education and training goals.

Massive Open Online Courses

Several data science courses and programs are burgeoning as online courses or MOOCs. These offerings come in many forms, including stand-alone courses and series of multiple linked courses. Some MOOCs can provide certificates to demonstrate completion of a course or series.

Coursera,[44] edX,[45] and others provide offerings in data science. For example, current data science offerings from edX include the following programs:

---

[38] The website for the University of Michigan certificate in data science is http://midas.umich.edu/certificate/, accessed February 18, 2018.

[39] For an online directory of graduate-level certificates in data science, see http://www.mastersindatascience.org/schools/certificates/, accessed February 12, 2018.

[40] The website for the Columbia University Data Science Institute is https://industry.datascience.columbia.edu/executive-training, accessed February 12, 2018.

[41] The website for the Kellogg Executive Education programs is http://www.kellogg.northwestern.edu/executive-education/individual-programs/executive-programs/bigdata.aspx, accessed February 12, 2018.

[42] The website for the edX Data Science for Executives program is https://www.edx.org/professional-certificate/data-science-executives, accessed February 12, 2018.

[43] The website for the Coursera Executive Data Science Specialization is https://www.coursera.org/specializations/executive-data-science, accessed February 12, 2018.

[44] The website for the Coursera data science offerings is https://www.coursera.org/browse/data-science?languages=en, accessed February 20, 2018.

[45] The website for the edX data science offerings is https://www.edx.org/course/subject/data-analysis-statistics/data-science-courses, accessed February 20, 2018.

- A Microsoft Professional Program in Data Science built on nine courses, starting from Excel, expanding to R and Python, and leading into a project.[46]
- A MicroMasters in data science by the University of California, San Diego, consists of four courses: Python for Data Science, Statistics and Probability for Data Science using Python, Machine Learning Fundamentals, and Big Data Analytics using Spark.[47]
- A MicroMasters in big data by the University of Adelaide consists of five courses focusing on programming, computational thinking, big data fundamentals and analytics, and a capstone project.[48]
- A version of the University of California, Berkeley, introductory course Data 8: Foundations of Data Science is offered as three 5-week courses forming a professional certificate: Computational Thinking with Python, Inferential Thinking by Resampling, and Prediction and Machine Learning.[49]

Current Coursera offerings include the following:

- A data science specialization, created by Johns Hopkins University with industry partnership from SwiftKey and Yelp, is made up of 10 courses designed to expose students to the full data science life cycle and concludes in a capstone project.[50]
- A specialization in probabilistic graphical models, created by Stanford University, requires a total of three 5-week courses in representation, inference, and learning.[51]
- A specialization in business statistics and analysis, created by Rice University with industry partner BaseMetrics, provides an overview of the tools and techniques used by data analysts in the business world via four courses and a capstone experience.[52]

---

[46] The website for the Microsoft Professional Program in Data Science is https://www.edx.org/microsoft-professional-program-data-science, accessed February 20, 2018.

[47] The website for the MicroMasters in data science is https://www.edx.org/micromasters/data-science, accessed February 20, 2018.

[48] The website for the MicroMasters in big data is https://www.edx.org/micromasters/big-data, accessed February 20, 2018.

[49] The website for the Data 8: Foundations of Data Science MOOC is https://www.edx.org/professional-certificate/berkeleyx-foundations-of-data-science, accessed February 20, 2018.

[50] The website for the Coursera data science specialization is https://www.coursera.org/specializations/jhu-data-science#about, accessed February 20, 2018.

[51] The website for the Coursera specialization in probabilistic graphical models is https://www.coursera.org/specializations/probabilistic-graphical-models, accessed February 14, 2018.

[52] The website for the Coursera specialization in business statistics and analysis is https://www.coursera.org/specializations/business-statistics-analysis, accessed February 14, 2018.

- A specialization in data analysis and interpretation, created by Wesleyan University in partnership with DRIVENDATA and The Connection, teaches the fundamentals of data science through four project-based courses, using either SAS or Python, and a capstone experience.[53]

MOOCs can be valuable resources for fast on-ramping, supplementing curriculum, and educating the citizen scientist in data science, albeit with low completion rates (Chuang and Ho, 2016). They can also provide content that can be utilized in traditional classroom settings. However, limitations exist. In order for MOOCs to succeed, they need to be organized, cohesive, thorough, and interactive. MOOC developers, like any curriculum developers, will also confront the challenge of keeping pace with quickly evolving data science topics.

Summer Programs and Boot Camps

Summer programs and boot camps provide students the opportunity for focused data science experiences, which can supplement their formal education and on-the-job training.

Summer Programs

Currently, there are relatively few data science-focused summer programs targeted toward undergraduate students—far fewer than those aimed at graduate students—but this is likely to change in the future as more undergraduate students are exposed to data science topics. The majority of these programs are organized by academic institutions and are designed to fast-track undergraduate students into data science by balancing experiential learning, projects with real data, and team building with some course instruction. Current programs tend to be small (from 8 to 40 students), which allows many to fully fund students through fellowships and scholarships. The programs are of varying durations, ranging from 1 week to 12 weeks.

Summer programs are often driven by a theme and target specific types of students—engineering/mathematics/computer science research-bound students (e.g., the Iowa State University Midwest Big Data Summer School[54]), underrepresented students from a broad spectrum of

---

[53] The website for the Coursera specialization in data analysis and interpretation is https://www.coursera.org/specializations/data-analysis, accessed February 14, 2018.

[54] The website for the Iowa State University Midwest Big Data Summer School is http://mbds.cs.iastate.edu/2017/, accessed February 12, 2018.

majors (e.g., the Microsoft Research Data Science Summer School in New York City[55]), students interested in societal impact (e.g., the University of Chicago Data Science for Social Good[56]), or students interested in health applications (e.g., the University of Michigan Big Data Summer Institute[57]). Testimonials from students in the longer-duration summer schools indicate that they benefit from the long-term immersive environment, where they have a chance to develop community, communication, and teamwork. These immersive programs can be life changing, altering students' academic and career pathways and often motivating them to go to graduate school.

Summer programs are an effective mechanism for student on-ramping that can energize students to develop a passion for data science. However, they are currently reaching only the relatively small number of students who are prepared for an intensive and sometimes stressful, fast-paced program. By their very nature, these programs work best when the small groups of students are together long enough to benefit from the immersive summer school experience, actively collaborating on learning, solving problems, and forming bonds of friendship. Scaling summer programs so that they reach a larger number of students faces several challenges. Developing a summer school demands a considerable amount of effort for faculty, requiring careful orchestration of a challenging quick-delivery curriculum accompanied by hands-on projects and other activities. There is scant availability of grant funding for development of the unique, multidisciplinary curriculum of data science, and there are not many funding sources for scholarships for students in need of financial aid. Furthermore, keeping the student-to-teacher ratio small will require additional resources for faculty and graduate teaching assistants that may not be available to academic institutions. There is evidence that programs can be more effective when they include a mix of highly motivated students that includes underrepresented minorities or economically disadvantaged students (Fine and Handelsman, 2010). An additional challenge is maintaining a healthy learning environment for students as they struggle to understand concepts in these fast-paced programs. Established academic and industry partnerships on data science summer programs are rare but would be worthwhile.

---

[55] The website for the Microsoft Research Data Science Summer School is https://ds3.research.microsoft.com/, accessed February 12, 2018.
[56] The website for the University of Chicago Data Science for Social Good program is https://dssg.uchicago.edu/, accessed February 12, 2018.
[57] The website for the University of Michigan Big Data Summer Institute is https://sph.umich.edu/bdsi/, accessed February 12, 2018.

Boot Camps

Many data science boot camps offered today are for-profit ventures that aim for professional development, corporate training, and continuing education (i.e., training and preparation of the data science workforce). They aim to teach large numbers of students. To do this, they frequently blend online and on-site learning, with the principal instructors online and local instructors on-site to reinforce training. The programs last from 1 weekend to 12 weeks and are organized into modules tightly coupled to workforce skills gleaned directly from industry. Many cover advanced topics like deep learning and recommender systems. These boot camps often have tracks that distinguish among subspecialties of data science, such as data engineering, data analytics, or data science for business.

Boot camps are often run by instructors who are practicing data scientists in industry. They can teach students to use a wider range of tools than students may experience in the classroom, and course content can quickly evolve as new tools emerge. Current tools include Pandas, BeautifulSoup, Seaborn, D3.js, R, and Python; however, data management, visualization, and analysis tools are continuously evolving. Boot camp participants may also have the opportunity to explore the data science life cycle in greater depth with student-driven projects: students may pose their own questions and collect their own data rather than starting projects after data have already been collected. With their deeper contacts with industry, boot camp instructors can offer direct links to the broader data science community. Many boot camps offer career advising, job placement, and networking opportunities as well.

Founded in 2013, Metis[58] is a boot camp with locations in New York, California, Illinois, and Washington. Metis offers 12-week boot camps that aim to bridge the gap between industry and academia and serve as a complement to other learning mechanisms. Participants learn a combination of theoretical concepts and applications including how to ask a solvable question, scope projects, collaborate and communicate with multidisciplinary groups, and use emerging tools and technologies.

Another example of a data science boot camp is that provided by General Assembly, with locations throughout the United States and abroad. Its 12-week immersive program in data science,[59] described as a "career accelerator," includes units on Git, UNIX, and relational databases; data analysis and Python; machine learning, modeling techniques,

---

[58] The website for Metis is https://www.thisismetis.com/data-science-bootcamps, accessed January 25, 2018.

[59] The website for General Assembly's immersive experiences is https://generalassemb.ly/education/data-science-immersive, accessed February 1, 2018.

and big data; critical thinking and synthesis; and visualization, presentation, and reporting.

DataCamp[60] offers a unique boot camp experience in that all of its coursework is provided in an online learning environment. Twenty different "tracks" are offered, depending upon whether the participant wants to develop particular skills or to train for a new career, or whether the user wants to develop expertise in R or Python, for example. A number of "project" courses have been created to help refine and integrate earlier knowledge.

While these boot camp experiences can be positive for students, they also tend to be very expensive and therefore out of reach for many undergraduate students. Many boot camps are specifically designed for recent graduates or professionals seeking a career change and thus may not be appropriate for undergraduate students. The immersive experience that a boot camp offers may also not be the right model for certain people, depending on their schedules, their career or educational goals, their learning styles, and their previous training (Feldon et al., 2017).

However, the unique position of these boot camps at the interface between education and industry could offer lessons for data science undergraduate-focused boot camps and on-ramping activities. Their ability to adjust in real time to industry demands, their intent to deliver sustainable fundamental data science skills, and their emphasis on project-based experiences would be applicable and beneficial in other settings and will continue to serve as a complementary activity to online courses, advanced degrees, and hackathons.

An alternative model to the mostly for-profit boot camps discussed above is the workshop approach that has been extensively developed by the nonprofit Carpentries[61] and given to thousands of participants worldwide. This organization has motivated a large number of students and researchers to receive training, initially focused on effective and quality software development through Software Carpentry and more recently on fundamental data skills through Data Carpentry. Their workshops are typically a few days in length and hosted by institutions and partners desiring a domain-specific introduction to basic data science for graduate students, faculty, and local researchers. The Carpentries accomplish this at low cost through volunteer instructors who complete an extensive training program. Although much of this enterprise has focused on more advanced students with explicit domain knowledge but little prior computational experience, the approach could be expanded to focus on undergraduates.

---

[60] The website for DataCamp is https://www.datacamp.com, accessed February 1, 2018.
[61] The website for the Carpentries is https://carpentries.org/, accessed April 23, 2018.

Finding 3.1: Undergraduate education in data science can be experienced in many forms. These include the following:

- Integrated introductory courses that can satisfy a general education requirement;
- A major in data science, including advanced skills, as the primary field of study;
- A minor or track in data science, where intermediate skills are connected to the major field of study;
- Two-year degrees and certificates;
- Other certificates, often requiring fewer courses than a major but more than a minor;
- Massive open online courses, which can engage large numbers of students at a variety of levels; and
- Summer programs and boot camps, which can serve to supplement academic or on-the-job training.

Recommendation 3.1: Four-year and two-year institutions should establish a forum for dialogue across institutions on all aspects of data science education, training, and workforce development.

MIDDLE AND HIGH SCHOOL EDUCATION

There is considerable potential to infuse data science education into middle school and high school curricula, particularly in laying the groundwork for many of the aspects of data acumen discussed in the previous chapter (Finzer, 2013). There is an opportunity for college-level courses to drive data science content down into middle and high school curricula. For example, the curriculum from Jevin West's and Carl Bergstrom's course at the University of Washington, Calling B.S.: Data Reasoning in a Digital World, is being adapted by and adopted in high school classrooms across the country (UW, 2017). High school teachers have found the curriculum to provide an innovative method to teach students how to analyze information responsibly. Having such experiences at the high school level better prepares students both for postsecondary curricula and for the data-driven workforce that awaits them. For example, the New York Hall of Science[62] provides an opportunity for children, young adults, and their families to increase their understanding of data science through various interactive museum exhibits, data fests, and mobile city science programs.

---

[62] The website for the New York Hall of Science is https://nysci.org/, accessed February 22, 2018.

Such programs appeal to many students and community members as they offer an engaging alternative to traditional classroom learning.

However, infusing data science into the middle and high school curriculum is not a simple task, especially with changes in mathematics education as a result of the Common Core State Standards.[63] Although, as of early 2018, 42 states and the District of Columbia have adopted their academic standards to be aligned with the Common Core (for some states with modifications), there are still some states that have not. This means that there is still variability across states in terms of the progression and coherence of learning opportunities that students are presented with for mathematics instruction. Teachers, particularly middle school educators, have expressed feeling overburdened with the standards as the development of curricular materials has lagged and guidance is still needed with respect to classroom implementation (Bay-Williams, Duffett, and Griffith, 2016). Whereas this could be viewed as a complication for a simple infusion of data science into at least the middle school classroom, it could also be presented as an opportunity given the need to generate materials instead of overhauling existing ones.

Course sequencing issues that are prevalent at the middle school level also exist at the high school level, although such issues can be heightened owing to limitations in the organizational structure of high school mathematics curricula. Specifically, some courses become "gatekeepers" to more rigorous mathematics courses, and students who are "tracked" through a particular sequence of courses may not be presented with the opportunity to develop strong postsecondary mathematical skills (Gamoran, 2009; Lucas, 1999; Lucas and Berends, 2002; Oakes, 2005). These issues can be compounded by issues of equity and access (Cha, 2015; Dondero and Muller, 2012; Lleras, 2008), which could have implications for students' access to *any* data science instruction in middle and high school. Therefore, careful consideration is needed to ensure that the infusion or placement of data science into the mathematics curriculum for middle and high school allows for equitable access and opportunities for a broad spectrum of students.

Interested educators at both the middle and high school levels would benefit from access to more resources that will allow them to integrate data science concepts into their classroom teaching, especially as curriculum materials are still being developed.[64] The data science community,

---

[63] The website for the Common Core State Standards Initiative is http://www.corestandards.org/standards-in-your-state/, accessed February 23, 2018.

[64] Dozens of high schools in California are already offering data science classes for eleventh and twelfth graders that combine statistics and programming instruction through hands-on activities. For more information, see Jones (2018).

perhaps through a future professional society (discussed in Chapter 5 of this report), has an opportunity to better engage middle and high school educators through national conferences and online information sharing mechanisms. Special consideration and outreach to schools with students from predominantly underrepresented backgrounds may allow for increased opportunities for access to data science concepts.

REFERENCES

Adhikari, A., and J. DeNero. 2018. *Computational and Inferential Thinking: The Foundations of Data Science.* https://www.inferentialthinking.com/. Accessed April 17, 2018.

Amherst College. 2017. "Data Science." https://www.amherst.edu/academiclife/departments/courses/1718F/STAT/STAT-231-1718F. Accessed January 25, 2018.

Bay-Williams, J., A. Duffett, and D. Griffith. 2016. "Common Core Math in the K-8 Classroom: Results from a National Teacher Survey." https://eric.ed.gov/?id=ED570138. Accessed March 29, 2018.

CCAC (Community College of Allegheny County). 2018. "Data Analytics Technology (788): Associate of Science." https://www.ccac.edu/Data_Analytics_Technology.aspx. Accessed March 29, 2018.

Cha, S.-H. 2015. Exploring disparities in taking high level math courses in public high schools. *KEDI Journal of Educational Policy* 12(1):3-17.

Chuang, I., and A. Ho. 2016. "HarvardX and MITx: Four Years of Open Online Courses—Fall 2012-Summer 2016." http://dx.doi.org/10.2139/ssrn.2889436. Accessed April 1, 2018.

Dondero, M., and C. Muller. 2012. School stratification in new and established Latino destinations. *Social Forces* 91(2):477-502.

Feldon, D.F., S. Jeong, J. Peugh, J. Roksa, C. Maahs-Fladung, A. Shenoy, and M. Oliva. 2017. Null effects of boot camps and short-format training for PhD students in life sciences. *Proceedings of the National Academy of Sciences* 114(37):9854-9858.

Fine, E., and J. Handelsman. 2010. "Benefits and Challenges of Diversity in Academic Settings." Brochure prepared for the Women in Science and Engineering Leadership Institute. http://wiseli.engr.wisc.edu/docs/Benefits_Challenges.pdf.

Finzer, W. 2013. The data science education dilemma. *Technology Innovations in Statistics Education* 7(2):1-9.

Gamoran, A. 2009. Tracking and inequality: New directions for research and practice. Pp. 213-228 in *The Routledge International Handbook of the Sociology of Education*, eds. M.W. Apple, S.J. Ball, and L.A. Gandin. New York: Routledge.

Jones, C. 2018. "Big data" classes a big hit in California high schools. *EdSource*, February 19. https://edsource.org/2018/big-data-classes-a-big-hit-in-california-high-schools/593838. Accessed March 22, 2018.

Lleras, C. 2008. Race, racial concentration, and the dynamics of educational inequality across urban and suburban schools. *American Educational Research Journal* 45(4):223-233.

Lucas, S.R. 1999. *Tracking Inequality: Stratification and Mobility in American High Schools.* New York: Teacher's College Press.

Lucas, S.R., and M. Berends. 2002. Race and track location in U.S. public schools. *Research in Social Stratification and Mobility* 25:169-187.

Montgomery College. 2018. "Data Science Certificate: 256." http://catalog.montgomerycollege.edu/preview_program.php?catoid=8&poid=1877&returnto=1322. Accessed January 25, 2018.

Nashua Community College. 2018. "Why Choose Foundations in Data Analytics?" http://www.nashuacc.edu/academics/associate-degrees/stem-and-advanced-manufacturing/398-foundations-in-data-analytics. Accessed March 29, 2018.

Oakes, J. 2005. *Keeping Track: How Schools Structure Inequality.* New Haven, Conn.: Yale University Press.

UC Berkeley (University of California, Berkeley). 2018. "Data 8: Foundations of Data Science." http://data8.org. Accessed January 25, 2018.

UC San Diego (University of California, San Diego). 2017. "Data Science Undergraduate Program." http://dsc.ucsd.edu. Accessed January 25, 2018.

UW (University of Washington). 2017. "Calling Bullshit" makes an impact at schools across the country. https://ischool.uw.edu/news/2017/10/calling-bullshit-makes-impact-schools-across-country. Accessed February 22, 2018.

4

Starting a Data Science Program

Starting a new data science program is challenging, to say the least. As with any new academic program, a curriculum needs to be determined, resources and faculty need to be found, and some means of assessment needs to be implemented. However, data science programs pose particular challenges owing to their interdisciplinary nature, the broad set of topics they encompass, and the acquisition of data and large-scale computational infrastructure they require.

Thus, launching a new undergraduate program in data science may be a significant undertaking in many institutions. Administrators and program developers will face myriad decisions. Should a new department be created to support this program? Or should existing departments take on the challenge solely or in collaboration with other departments? What content/level of high school knowledge will be useful or required of students entering the data science program? How will data science be integrated into the curriculum? Should it be included at the very beginning of a student's coursework, after some prerequisite coursework, or as a capstone? Will it be a major, a minor, a general education requirement, or all of the above? Which mathematics, statistics, and computer science courses should be required of data science students? Should these be taught as separate courses, or should the content be integrated? How might institutions appropriately utilize the collections of online resources available and "downscale" these to appropriate levels if they are focused on more advanced training? Institutions developing undergraduate programs will also need to consider how ethics and communication will be included in the curriculum, as well as how to ensure that the program is accessible to students from varied backgrounds.

New data science programs require resources, broad discussions with faculty and leadership across the institution, and perhaps approval through formal bodies. The backing of the administration as well as broad support from multiple departments is typically necessary, and attention to the costs and funding model from the outset can greatly increase the chance of success.

As institutions examine how best to provide data science education to their students, one solution may be to reconstitute, combine, or reenvision already existing curricula. Much of the research on how best to teach science, technology, engineering, and mathematics (STEM) concepts will be readily applicable. (See the discussion of data acumen attributes in Chapter 2 of this report for examples of introductory and advanced concepts.) While some coursework could be immediately swapped into a data science program, it is likely that this will take more forethought and planning to appropriately consider the learning outcomes and content knowledge that data science students need to have. In a number of programs (see Chapter 3), the first official data science offering is a brand-new class, meant to serve as a rich introduction to what it means to practice data science. In institutions with less funding or expertise for course development, the need to get a program up and running may push toward more borrowing of content if not whole courses. However, a strong data science program is likely to need eventually to move beyond "patching together" a curriculum or class.

In this section, the committee describes the key challenges that academic institutions will face as they set up a program. But this section begins with an opportunity: an important element of program design is ensuring that the program is welcoming and inclusive to all students, regardless of their identity-related characteristics or educational background and attainment.

ENSURING BROAD PARTICIPATION

According to the South Big Data Innovation Hub's *Keeping Data Science Broad*, "the variety of perspectives such diversity [in terms of race, gender, religious affiliation, socioeconomic status, ethnicity, and first-generation status] provides is as essential as that provided by the transdisciplinary nature of data science for innovation and growth of the field" (Rawlings-Goss, 2018, p. 29). The report explains that the first step in creating a more inclusive environment is to ensure that students and faculty alike—at *all* types of educational institutions—have *equitable* access to resources (e.g., high-quality data, tools, technology, adaptable and appropriate curriculum, and advisors). Also crucial to retaining broad participation in data science is a "culturally relevant curriculum," a more diverse faculty, and collaborations between majority-serving and minority-serving institutions (Rawlings-Goss, 2018, p. 31).

Thus, it is the responsibility of academic institutions to ensure inclusion and broad participation and engagement in data science programs. Master (2017) suggests that data science programs at higher education institutions increase exposure to data science fields, broaden beliefs about who belongs in these fields, challenge students' beliefs about fixed abilities, and show that data science can make a difference in society in order to broaden participation and engagement in data science. Williams (2017) suggests that faculty adjust curriculum to be more inclusive, create opportunities for students to engage in community data, affirm student ability, and create diverse teams of students. The efforts highlighted by Master and Williams not only lead to increased engagement, but they also stand to *sustain* participation of underrepresented populations in data science. If data science is to avoid a similar decrease in participation that occurred in the 1980s in computer science among female students, it is imperative that underrepresented students are supported both academically and through mentorship, recognizing the opportunities that the field of data science presents and the value they can add to it.

Some of the introductory data science courses described in this report have made inclusion and broad participation a central goal, shaping pedagogy, technical infrastructure, and staffing. Some notable steps include the following:

- Designing the material to avoid the need for mathematics, statistics, or programming prerequisites beyond that required for entry to the academic institution, thereby avoiding demographic skews that such prerequisites might induce.
- Using a computing infrastructure that does not rely on personal laptops or access to computer labs; possibly hosting the infrastructure entirely in the cloud so that it can be accessed through a web browser.
- Providing teams of laboratory assistants and tutors to give additional support for students needing assistance.
- Choosing project topics carefully to be of broadest interest and to raise awareness of social issues.
- Operating a cohort-based "data scholars" program[1] in concert with the instructional program to address issues of underrepresentation.

---

[1] The website for the University of California, Berkeley, Data Scholars Program is https://data.berkeley.edu/education/data-scholars, accessed February 20, 2018.

Additionally, the huge opportunity of data science to be a gateway to STEM careers should be emphasized. The wide range of applications of data science to multiple fields, including humanities, social sciences, and the arts, expands the reach of STEM into society. Couching data science in terms of a life skill and a cultural pursuit can help reshape the image of science and increase the number of students interested in STEM fields. Therefore, use cases should be drawn not just from other STEM or scientific disciplines; they should also be drawn heavily from the arts, humanities, social sciences, and popular culture to attract new entrants to the field.

As many data science programs are being freshly created, ample opportunity exists to build broad participation from the beginning. While not a panacea, there are several actions that data science programs can take to broaden participation. The Joint Working Group on Improving Underrepresented Minority Persistence in STEM offered the following recommendations for broadening participation of underrepresented minorities in STEM programs (Estrada et al., 2016):

- Track and increase awareness of institutional progress toward diversifying STEM;
- Create strategic partnerships with programs that create lift;
- Unleash the power of the curriculum and active learning;
- Address student resource disparities; and
- Stimulate students' creativity.

The Joint Working Group points out that there are many programs that have been successful in attracting and retaining underrepresented students in STEM disciplines. For example, in 2017, over half of the computer science graduates from Harvey Mudd College were women (Williams, 2017). Harvey Mudd has succeeded in attracting and retaining underrepresented students in part owing to its commitment to fostering a "growth mindset" in these students (Dweck, 2006). Harvey Mudd faculty teach problem solving using real-world examples, offer four unique styles of one introductory computer science course based on student knowledge and interest, require students to work together in completing homework assignments, and actively encourage students to enroll in a subsequent computer science course (Williams, 2017). To engage potential future students in STEM, Harvey Mudd also hosts a conference at which African American and Hispanic middle and high school girls have the opportunity to build partnerships with professional women practicing in STEM fields. Data science programs may benefit from looking to these and other successful STEM initiatives as models for attracting and retaining underrepresented students.

The Joint Working Group also suggests that such programs are most effective when coupled with assessment and evaluation data that "clearly show the amount of progress or disparity that exists at the institutional level" (Estrada et al., 2016, p. 8). Another way to increase participation is to avoid filter or gate-keeping courses (especially early in the program) and replace them with courses that entice student participation through heightening the excitement and applicability of data science. It may also behoove data science programs to consider which faculty are teaching first-year or introductory data science courses, ensuring that these faculty members can connect with and engage students.

Data science programs also need to embrace multiple entrance points into the discipline—think of the metaphor of a "watershed" in which students from a variety of educational backgrounds and fields can enter, rather than a "pipeline" from one or more particular fields into data science. Varma (2006) agrees that the notion of the pipeline does not extend far enough, as underrepresented minority students face heightened entry and retention barriers. In combination with the recommendations from the Joint Working Group, increasing teacher assistant training, awareness for advising staff, communication between students and faculty, and partnerships with high school teachers could help postsecondary data science programs retain a more diverse student body (Varma, 2006). Curricular options such as minors or data science add-ons to substantive disciplines are two possible ways to open the data science enrollments. In addition to focusing on programs in science fields and broadening participation there, programs in popular areas such as music or communications could be targeted for outreach.

> **Finding 4.1:** The nature of data science is such that it offers multiple pathways for students of different backgrounds to engage at levels ranging from basic to expert.
>
> **Finding 4.2:** Data science would particularly benefit from broad participation by underrepresented minorities because of the many applications to problems of interest to diverse populations.
>
> **Recommendation 4.1: As data science programs develop, they should focus on attracting students with varied backgrounds and degrees of preparation and preparing them for success in a variety of careers.**

ACADEMIC INFRASTRUCTURE

The popularity of data science courses and programs will affect academic infrastructure in several ways—notably, in terms of who will "own" the program and how it will be delivered. Faculty and administrators will need to examine how the goals of data science education align with the institution's current infrastructure. What departments and colleges should be involved? Because data science intersects with mathematics, statistics, computer science, and other domains, institutions need to consider whether data science needs to become a stand-alone department or be integrated with other departments. Administrators will need to consider ways to motivate departments to work with one another across disciplines, and department chairs will need to consider ways to motivate their faculty to participate in the implementation of innovative new curricula, whether or not they are in the "home" department. This holistic approach toward data science education is crucial, particularly given the interdisciplinary nature of the field of data science.

Furthermore, given its interdisciplinary nature, a new data science program at the undergraduate level needs to involve the collaboration of several disciplines and programs. However, few instructors are likely to be available who are equally able to teach classes in the full complement of fields. Initially, at least, creative ways of involving faculty from multiple departments is likely to be necessary, so that they can learn from each other and so that students get the broad view of data science that the committee envisions.

However, cross-departmental or institutional collaboration to develop data science programs may prove easier in theory than in practice. In some colleges and universities, academic tribalism and the increased importance of tuition generation might impede these programs from being truly interdisciplinary. Thus, the flexibility to hire or train faculty in the multiple aspects of data science will be necessary to ensure that all programs still achieve their educational goals.

As one example, consider Virginia Tech's solution to the organizational model. Virginia Tech offers a major in computational modeling and data analytics. The departments that host the major (i.e., computer science, statistics, and mathematics) span two colleges (i.e., the College of Engineering and the College of Science), making interdisciplinary communication and cooperation extremely important. To foster productive collaboration among and within its five interdisciplinary programs, the College of Science set up the Academy of Integrated Science, which is a department-level organizational structure that helps interdisciplinary programs by managing budgets, undergraduate advising, student recruitment, and assessment. Having such a body in place allows faculty to focus solely on developing and delivering curriculum to the students. The

Academy of Integrated Science also develops a memorandum of understanding for new faculty hires that establishes their roles in both their home departments and in the interdisciplinary programs (Embree, 2017).

Such cross-departmental collaboration requires new mechanisms for both funding and encouragement. Opportunities for a wide variety of faculty to participate in data science programs will need to be created, as will incentives and rewards for those faculty teaching data science. Reward systems more generally may need to adjust to place greater value on teaching more students, especially when that means there will be greater diversity in their level of preparation. As data science begins to enter conversations in many disciplines, educators and administrators will have to consider the roles of the humanities, social sciences, and arts programs. There are also opportunities for developing programs for students in non-STEM fields, although there are risks that these become "data science-lite" programs that add limited marketable or intellectual value to students.

Several specific hurdles to launching and sustaining data science programs have been encountered and to some extent overcome at various academic institutions. Some of these challenges are associated with growing pains of starting up any new program that is in high demand:

- *Overcoming initial resistance.* One of the first challenges prospective programs have to overcome is initial resistance by established departments and programs to launching a new program that is in intellectual proximity and competes for tuition dollars and other resources. This is especially challenging for data science, as it has a large footprint across many professional, scientific, and engineering disciplines.
- *Recruiting and retaining faculty.* Another important challenge is recruiting faculty to create and teach integrative introductory courses in data science and to serve as advisors and mentors for data science students. Departmentally centered tenure and promotion criteria may lead junior faculty to be reluctant to devote much time to launching new programs. An additional challenge has been retention of data science faculty in an economic environment where faculty are increasingly lured away by industry.
- *Developing curricula.* It is often challenging to develop a consensus on a core curriculum that best serves the various interests and backgrounds of data science students. In this era where many of the existing data science-related courses are oversubscribed, other departments can be reluctant to enroll data science students in their popular courses (e.g., machine learning, data mining, natural language, applied statistics) because doing so may take seats away from their own students.

- *Providing physical space.* To be most effective, data science programs need flexible physical space to create the collaborative environment in which their students thrive. Such well-situated space is often scarce.
- *Facilitating interactive experiences.* There is a lack of sustainable and scalable models for capstone programs and similar experiential integrative experiences that have been shown by the Association of American Colleges and Universities (2013) and others to be high-impact educational practices.
- *Encouraging industry partnerships.* With high turnover in the industry workforce, colleges are facing the challenge of building lasting industry relationships to keep education and training well matched to the needs of the rapidly evolving data science workforce.

Additional infrastructure considerations include enrollment budgets, strategies to build a data science major curriculum (i.e., prerequisites, introductory, advanced, applied, capstone), and ways to align general education requirements with data science. Institutions will need to consider how to provide and share resources for their varied data science experiences (e.g., textbooks, teaching materials, open access, clearinghouse). Advising will also be important for the success of data science undergraduate programs. Formal evaluation methods will need to be implemented to gauge the success of these programs and improve them. The Moore-Sloan Data Science Environments (2018) have put forth some suggestions on creating institutional change in data science, including establishing a neutral space for students and faculty to gather, providing access to professional data scientists and research software engineers who can assist and serve as role models, developing a data science consulting capability, considering the scalability of data science educational initiatives, encouraging software and data openness and reuse, and involving a wide range of people in data-intensive discovery.

Another challenge will be that many fields involved with data science are themselves experiencing rapid change and evolution. As a consequence, data science curricula will also likely evolve rapidly, and programs need to be ready and willing to adapt. This will undoubtedly lead to the same types of questions that have been explored in computer science and other rapidly evolving fields in past years. Last, it behooves institutions to consider the alternative pathways students might take into data science by removing obstacles and barriers for students who want to change their concentration to data science during the course of their studies or making it easier to add a data science minor. Overcoming these challenges will require institutions to broker between competing interests, to recruit new faculty and staff in data science, and to make strategic long-term investments to sustain the activity.

Finding 4.3: Institutional flexibility will involve the development of curricula that take advantage of current course availability and will potentially be constrained by the availability of teaching expertise. Whatever organizational or infrastructure model is adopted, incentives are needed to encourage faculty participation and to overcome barriers.

Computational Infrastructure

A major driver of data science education has been the evolution of data and the infrastructure for accessing it and analyzing it. Hands-on experience with the entire data science life cycle is an essential part of the training and education of data science students, regardless of the educational modality. In particular, students need to be taught how to handle large amounts of data and how to run scalable but sophisticated analysis software on the data—often requiring distributed data storage, multi-core processing, and parallel computation. However, maintaining large complex data sets and high-performance computing systems on college campuses strains the resources of educational institutions. While several of the larger research universities retain high-performance computing and large server facilities, most universities and colleges are in the process of transitioning their computing and storage to cloud service providers that provide students reliable access to their data and the computational resources to run algorithms against the data. Thus, the cloud has played and will continue to play an important role in transforming data science education. A logical next step might be for colleges to band together to federate these cloud resources under an "academic cloud."[2] Such a federated academic cloud could provide common platforms for students across the nation, facilitating data integration and analysis, reducing costs to educational institutions, and balancing inequities in access to instructional resources.

Finding 4.4: The economics of developing programs has recently changed with the shift to cloud-based approaches and platforms.

CURRICULUM

As discussed in Chapter 2 of this report, there is a progression of topics and skill sets that will guide students to develop data acumen. Key concepts required to develop data acumen include mathematical

---

[2] The National Science Foundation, for example, invested $20 million in academic cloud computing in 2014 (Boland, 2014).

foundations, computational foundations, statistical foundations, data management and curation, data description and visualization, data modeling and assessment, workflow and reproducibility, communication, domain-specific considerations, and ethical problem solving. These skills then become transferable into a range of data science positions in the workplace.

Each undergraduate modality discussed in Chapter 3 offers a unique pathway to various data science careers. The degree to which each concept or skill is emphasized in each modality depends upon the respective career trajectories. While a 4-year data science degree may be most appropriate for some data scientists, a 2-year associate's degree may be better suited for others. And while a boot camp may help prepare a business professional to incorporate data analytics in the workplace, a data science minor may offer valuable training for data-driven decision makers in a variety of fields. It is important to note that, as the field of data science continues to evolve at a rapid pace, it will often be necessary to reevaluate the types of careers utilizing data science as well as the data science skill sets necessary to achieve success in those careers.

FACULTY RESOURCES

Mirroring the variety of pathways for data science education discussed in Chapter 3, there are a number of ways in which data science courses may be taught. Some data science courses, owing to their interdisciplinary nature, are taught either by a team of faculty or by two faculty with the appropriate areas of expertise to cover multiple perspectives. Though this approach offers the most well-rounded experience for students, it can be difficult to find the administrative support and additional resources needed. It remains challenging to recruit appropriate new faculty to teach both introductory data science courses and courses for the data science major or minor. Faculty need to have multiple experiences with data science projects to develop the perspective to guide their students. These faculty need to be diverse and have the ability to serve as role models for future data scientists, while also meeting competencies in the practical data acumen areas discussed in the previous section and in Chapter 2 of this report. As the field of data science expands, faculty are likely to be needed in an even broader range of competencies.

Considerations for current faculty are also necessary, as many will need to be retrained in new data science methods and tools, both of which will continue to evolve rapidly in the coming years. Faculty will also benefit from professional development in new teaching approaches to best meet the needs, learning styles, and knowledge levels of future undergraduate students. Such training will be especially useful for faculty teaching introductory classes composed of students with various academic backgrounds and career interests. Funded by the National Science Foundation, Training a New Generation of Statistics Educators[3] is an example of a program that creates professional learning communities whose members participate in workshops, mentorship programs, and national conferences, all in an effort to increase their statistical content knowledge and improve their teaching. The current program includes over 70 instructors from 2-year institutions across the United States (Posner, 2017). Some academic institutions have developed their own focused data science education programs for their faculty; for example, the University of California, Berkeley, offers summer workshops on the pedagogy and practice of data science to engage faculty across the university.[4]

Perhaps most challenging is retaining faculty in data science programs. Given their skill sets, professors of data science domains are highly sought after throughout industry (Kaminski and Geisler, 2012). While academic institutions are unlikely to have the resources to offer comparable salaries, they need to consider alternative incentives (e.g., opportunities for transdisciplinary collaboration, for pursuing open-ended research topics, for developing curricula, and for professional stability [e.g., tenure]) that will appeal to, entice, and retain data science faculty.

ASSESSMENT

In order for data science programs to flourish, the progress of data science students needs to be assessed early and often. Jordan (2017) asserts that assessment cannot be conducted without a clear understanding of the core skills for data science—collection, analysis, visualization, and sharing—as well as a clear definition of student learning outcomes (perhaps inspired by Bloom's Taxonomy). She suggests the following eight steps to create and assess the data science education classroom:

1. Understand the audience to create an impactful and positive learning environment,
2. Know what motivates the students,
3. Develop a code of conduct,
4. Create challenge questions and exercises,
5. Incorporate qualitative mechanisms to improve teaching and quantitative mechanisms to gauge student learning,

---

[3] The website for this National Science Foundation-supported project is https://www.nsf.gov/awardsearch/showAward?AWD_ID=1432251, accessed February 6, 2018.

[4] An overview of a collection of pedagogy workshops can be found at https://data.berkeley.edu/news/2018-data-science-education-opportunities, accessed April 22, 2018.

6. Design interventions around learning outcomes as well as around students' needs,
7. Conduct long-term follow-up, and
8. Continue to build the data science community.

REFERENCES

Association of American Colleges and Universities. 2013. Capstones and integrated learning. *Peer Review* 15(4).

Boland, R. 2014. NSF invests millions in academic cloud computing testbeds. *Signal*, August 21. https://www.afcea.org/content/nsf-invests-millions-academic-cloud-computing-testbeds. Accessed February 22, 2018.

Dweck, C. 2006. *Mindset: The New Psychology of Success*. New York: Ballantine Books.

Embree, M. 2017. "Forging Virginia Tech's CMDA Major Across Departments." Webinar Presentation to the Committee on Envisioning the Data Science Discipline: The Undergraduate Perspective, October 10. http://www.nas.edu/envisioningDS. Accessed February 14, 2018.

Estrada, M., M. Burnett, A.G. Campbell, P.B. Campbell, W.F. Denetclaw, C. Gutiérrez, S. Hurtado, et al. 2016. Improving underrepresented minority student persistence in STEM. *CBE Life Sciences Education* 15(3):es5.

Jordan, K. 2017. "Assessing Data Science Learning Outcomes." Webinar Presentation to the Committee on Envisioning the Data Science Discipline: The Undergraduate Perspective, October 24. http://www.nas.edu/envisioningDS. Accessed February 14, 2018.

Kaminski, D., and C. Geisler. 2012. Survival analysis of faculty retention in science and engineering by gender. *Science* 335(6070):864-866.

Master, A. 2017. "Diversity, Inclusion, and Increasing Participation in Data Science." Webinar Presentation to the Committee on Envisioning the Data Science Discipline: The Undergraduate Perspective, November 7. http://www.nas.edu/envisioningDS. Accessed February 14, 2018.

Moore-Sloan Data Science Environments. 2018. "Creating Institutional Change in Data Science." White paper. http://msdse.org/files/Creating_Institutional_Change.pdf.

Posner, M. 2017. "Go to the People: Impactful Faculty Training in Data Science." Webinar Presentation to the Committee on Envisioning the Data Science Discipline: The Undergraduate Perspective, September 26. http://www.nas.edu/envisioningDS. Accessed February 14, 2018.

Rawlings-Goss, R. 2018. *Keeping Data Science Broad: Negotiating the Digital and Data Divide Among Higher Education Institutions*. South Big Data Innovation Hub. http://bit.ly/KeepingDataScienceBroad_Report. Accessed March 28, 2018.

Varma, R. 2006. Making computer science minority-friendly: Computer science programs neglect diverse student needs. *Communications of the ACM* 49(2):129-134.

Williams, T. 2017. "Diversity and Inclusion in Data Science: Using Data-Informed Decisions to Drive Student Success." Webinar Presentation to the Committee on Envisioning the Data Science Discipline: The Undergraduate Perspective, November 7. http://www.nas.edu/envisioningDS. Accessed February 14, 2018.

5

Evolution and Evaluation

The variety of programmatic and curricular approaches described in the previous chapters points to numerous alternatives for development of data science learning in undergraduate programs. There are many different modalities for institutions to offer in data science education, as there have been in computer science and information science. Data science programs will continue to evolve within institutions and across the United States as driving factors, including student perception of career opportunities and funding for training programs from industry and government, modify student demand and associated institutional response.[1] As befits a new and evolving discipline, pilot programs designed to address different needs will arise at different types of institutions.

Evaluation of a program relies upon assessing student learning and gauging how well the program meets market needs. Evaluations can be used to test what works for an institution, given relevant constraints, and suggest when these pilot programs might be generalized for other contexts. It would be beneficial to use data science to continuously evaluate and evolve data science education (see Gertler et al., 2016). In particular, it would be useful to examine the impacts that different drivers of undergraduate programs have on the development of programs and their evolution. In this way, the programs would not evolve stochastically based upon particularities of the histories and personnel associated with

---

[1] See, for example, the Institute of Education Sciences What Works Clearinghouse, https://ies.ed.gov/ncee/wwc/, accessed February 20, 2018.

the programs, but rather they would innovate based upon data and the drivers at the institution.

The continued evolution of programs would benefit from an efficient means to collect, compare, and share evaluation data. Such capabilities could be established through collaborative efforts across a number of the professional societies that operate in the data science space or possibly from subsocieties that cross the variety of disciplinary boundaries in data science as well as the variety of application domains. Such collaborations could enhance the effectiveness and efficiency of educational programs by assisting in propagating methods for success. Eventually, formal schemes for standards, and possibly accreditation, could emerge from the application of data science methods to the variety of programs that have evolved.

EVOLUTION

There are multiple pathways to developing programs in data science, each with its own challenges and opportunities. While other domains have started by implementing undergraduate programs and expanding upward to include master's degree programs, or with doctoral programs and expanding downward, data science has taken an unusual path to date, as many master's degree programs have begun to be offered before undergraduate or doctoral programs. This institutional structure is perhaps driven by the rapidly arising industry demand for data scientists. This starting point influences how competencies are introduced into the curriculum. However, as Chapter 3 made clear, today different schools are developing multiple pathways from multiple starting points. This flexibility is beneficial, as the multitude of educational pathways enables students to be equipped to fill various future data science positions.

A 2015 National Science Board study described the importance of building a better understanding of the pathways taken by individuals to become part of the science, technology, engineering, and mathematics (STEM) workforce. It pointed out the need to "monitor and assess the condition of workforce pathways" as well as to "address roadblocks to the participation of groups traditionally underrepresented in STEM" (National Science Board, 2015). Reaching both of those goals requires building a model of the watershed of STEM-trained individuals and empirically assessing demand for STEM-trained employees (Metcalf, 2010).

In this section, the committee considers how data science programs might evolve. That evolution will be driven by a broad range of factors, including the starting point, the students who are the target of the program, and the institution that is developing that program. This section focuses on the maturation process and on ensuring program success.

The Maturation Process: Key Drivers of Change

Once started, programs will continue to evolve differently, based on the constraints of personnel and resources at the institution, student perceptions of careers in the area, and external influences such as industry and government partnerships. The long-term goal of any undergraduate educational program in data science is to produce successful data science students. But what constitutes success? Obtaining a degree? Gaining satisfactory employment (i.e., a job that is monetarily or intellectually rewarding)? Securing a place in graduate school? The answers to these questions will be driven by the institution as well as by the evolution of the workforce and the economy. If jobs are the students' goals, and the institution sees itself as preparing them for the job market, then industry's needs will be paramount, and undergraduate programs will need to evolve to meet those needs as the needs themselves evolve. Likewise, the availability of graduate programs (and their evolution over time) can also change the content and learning outcomes of undergraduate programs, which strive to ensure that their students have the right background to be admitted to these graduate programs.

Many subtler influences based on educational style may also be expected. Some institutions will tend to stress an understanding of theoretical foundations; other programs will focus on hands-on experience. The size of the school and popularity of the program will impact the types of opportunities available and the pedagogical style that may be appropriate. Over time, any of these may change, resulting in changes to individual courses and overall programs.

Since data science programs in general are new, there is little in terms of comprehensive data on general expectations for the undergraduates who enter them. Many programs that exist today were started as professional master's degree programs. Starting at the master's level presupposes knowledge and classes, usually with a specific disciplinary focus. There are also multiple possible pathways to data science education at the doctoral level, perhaps a Ph.D. in data science or a Ph.D. in a domain area with a specialization in data science. Many students entering such programs may not have the necessary prerequisites, so it is important for institutions to consider how to upskill and thus prepare them quickly.

Programs certainly will have more tactical intermediate goals, such as simply attracting more students to the program, broadening participation, or increasing the understanding of data science on campus. Another goal might be increasing experiential learning in data science through coursework or internships. Progress toward these goals can be measured and serve to pave the way to success—the percentage of various populations can be tracked, and delivery methods, classroom practices, or course content can be modified to attract a more diverse population as needed.

For example, with support from the National Science Foundation, the College Board developed a new advanced placement course, Computer Science Principles. This course is specifically designed to engage more high school students in computer science, and thus better prepare them for future study of STEM disciplines, by demonstrating the power of computer science to solve real-world problems through the use of a variety of computing tools and languages (NSF, 2014). Following the development of this new advanced placement course, pilot programs began at universities across the country (e.g., University of Washington[2] and North Carolina State University) to develop and offer a new introductory computer science curriculum, The Beauty and Joy of Computing. This curriculum is targeted toward non-computer science majors and centers around programming (using Snap!) and creating (UC Berkeley, 2018). Although such initiatives can be beneficial for students in affluent communities whose schools have the resources to offer advanced placement curricula, students from high schools without advanced placement curricula in mathematics or science disciplines will enter their undergraduate experience at a significant disadvantage from some of their peers, thus widening the knowledge gap that currently exists across student populations. It is crucial that funding organizations and curriculum developers continue to consider approaches that will best prepare all students for potential future work in data science. Other more widely accessible initiatives to attract diverse populations include efforts by Code.org to increase access to computer science in schools, especially for women and underrepresented minorities. The data collected from these and similar efforts, coupled with formal evaluation protocols (see the section "Evaluation," later in this chapter), can also guide the program evolution until the learning outcomes required to achieve success, however defined by a given program, are achieved.

Finding 5.1: The evolution of data science programs at a particular institution will depend on the particular institution's pedagogical style and the students' backgrounds and goals, as well as the requirements of the job market and graduate schools.

Pathways to Maturation

Given these various drivers, how will programs evolve? Again, there will be multiple pathways, but maturation will occur in both individual courses and at the program level. As pathways are developed, attention

---

[2] The webpage for Computer Science Principles at the University of Washington is https://courses.cs.washington.edu/courses/cse120/, accessed March 29, 2018.

will be needed to ensure that faculty and students are upskilled, flexibility is preserved and supported, and efforts are sustained. As instructors rework individual classes based on outcomes and evaluation, it is likely that they will replace borrowed content from existing courses with original materials that fit together more naturally and better match personal educational styles or the culture of that institution or department. Materials will be tuned to meet the objectives and the abilities of the students. New units, emphases, or experiences may be added (e.g., to teach a broader range of skills).

At the program level, there is likely to be an evolution in the courses over time, as well as in the responsibility or ownership of individual courses or even entire programs. Today, data science programs often originate in one existing department—frequently statistics, computer science, mathematics, or business. However, a program, although "owned" by one department, may well be composed of classes from multiple departments. Over time, that set of classes is likely to evolve as more custom data science classes are created. In some cases, an existing class from one department may be adopted as part of another department's data science curriculum but may eventually be duplicated (perhaps with some tailoring) to be offered by both departments; alternatively, a course may be shifted from one department to another altogether.

> **Recommendation 5.1: Because these are early days for undergraduate data science education, academic institutions should be prepared to evolve programs over time. They should create and maintain the flexibility and incentives to facilitate the sharing of courses, materials, and faculty among departments and programs.**

Ensuring Success

With the many ways that programs can evolve, academic and career advising are key components of a successful program. These are essential to sustain the viability of programs and pathways. Prior to launching a full-fledged program, advisors will need to be trained to understand the program content, anticipated pathways, and outcomes. This cadre of advisors will include not only those specific to the new program but also advisors for existing programs who will need to field new questions and be able to position the new program for success. How is data science different from statistics? It is not, for example, computer science with a focus on machine learning. It is not applied mathematics. It is a truly interdisciplinary activity that brings together domain scientists with computer scientists, information scientists, and statisticians. What is the best way for students to meet their goals? If they are unsure of their goals, how best can they explore the alternatives?

While institutions consider the many pathways available for their students, they will also need to consider who will develop and teach the general education and data science classes: teaching faculty, adjunct instructors, teaching assistants, or tenure track faculty. It is imperative that academic institutions have a balanced faculty with both domain knowledge and data science expertise so as to best deliver data science education that will prepare students for the varied and multidisciplinary workforce that lies ahead of them. As data science pedagogy teaches the appropriate skill sets, it may be wise for academic institutions to build up their graduate teaching assistantship programs so as to prepare more future data science educators. Both those who build the new curricula and those who teach it will need to be supported (i.e., motivated, trained, and rewarded) and upskilled in a variety of ways. More work will be needed to "teach the teachers" the multidisciplinary fundamentals of data science. Faculty could benefit from shared teaching materials (e.g., course notes, software, data, and case studies), as well as short courses and boot camps adapted to non-data scientists who will teach data science concepts. Course and curriculum builders could benefit from a roadmap on how to build curriculum and programs at all undergraduate levels, including what works (e.g., team teaching, student-initiated groups, teaching in context) and what does not.

Institutions that want to increase data science opportunities could consider the examples and lessons of interdisciplinary programs at other institutions, as well as how these may be able to provide guidance on how data science programs might evolve. Programs can benefit greatly from sharing ideas and data with each other, given the rapid emergence and evolution of this field. For example, the availability and usability (under a Creative Commons license) of the well-packaged curriculum and materials from the University of California, Berkeley, course Data 8: Foundations of Data Science (discussed in Chapter 3) has enabled several universities to quickly roll out similar courses to benefit their students.

Finding 5.2: There is a need for broadening the perspective of faculty who are trained in particular areas of data science to be knowledgeable of the breadth of approaches to data science so that they can more effectively educate students at all levels.

Recommendation 5.2: During the development of data science programs, institutions should provide support so that the faculty can become more cognizant of the varied aspects of data science

through discussion, co-teaching, sharing of materials, short courses, and other forms of training.

Finding 5.3: The data science community would benefit from the creation of websites and journals that document and make available best practices, curricula, education research findings, and other materials related to undergraduate data science education.

EVALUATION

Regardless of what educational approach is taken, it is important that rigorous evaluation of the outcomes is conducted so that the field can evolve in response to data and evidence. This evaluation is needed both at the level of individual programs as well as at the broader national level. The former could enable individual schools to continue to evolve their programs to better serve their students. The latter could enable the community to learn what works across many programs and eventually to evolve toward a common set of pathways and curricula that optimize outcomes for their students. This could also foster the development of educational research for this highly interdisciplinary field to provide a basis for determining the effectiveness of alternative approaches.

Figure 5.1 provides a simplified conceptual model that could underlie a process of evaluation at several hierarchical education levels. Obtaining and utilizing appropriate data at each step can help iteratively refine programs. Objectives, associated outcomes, metrics of success in reaching these outcomes, and the activities that impact these outcomes might be established at the level of a single program, at the level of an institution with multiple routes to success for data science students through various majors or minors, or at broader national scales.

Evaluation at the Level of an Individual Program

In developing data science pathways that align with their capabilities, academic institutions need to think about how these pathways can be evaluated. Discipline-based education research, which "investigates learning and teaching in a [science or engineering] discipline using a range of methods with deep grounding in the discipline's priorities, worldview, knowledge, and practices . . . [and] is informed by and complementary to more general research on human learning and cognition" (NRC, 2012), can be used to inform both evaluation and revision of data science curricula to best meet student needs. Both computer science and statistics, for example, have robust education communities whose results may be leveraged to improve data science education. However, further interdisci-

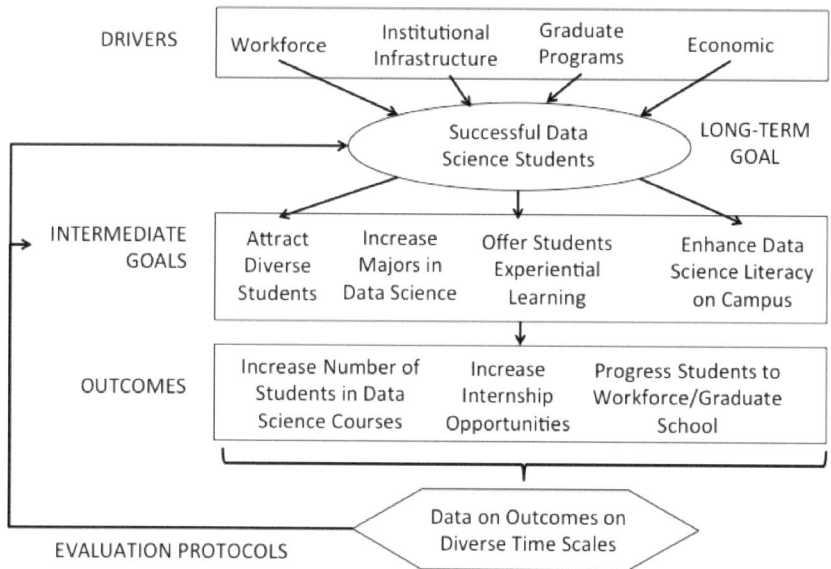

FIGURE 5.1 Simplified example theory of change model for undergraduate data science programs.

plinary education research is essential, given the interdisciplinary nature of data science and the related challenges of developing and evaluating such programs. As data science programs evolve, it is reasonable to think about how these programs could be accredited. However, it is premature for this committee to prescribe if or how accreditation should be done.

Evaluation using administrative records[3] has a long and distinguished history in education (Aaronson, Barrow, and Sander, 2007; see Box 5.1). Academic institutions with data science programs may consider matching their administrative records to U.S. Census Bureau or state wage record data (see Box 5.2) to track the outcomes of students trained in data science relative to other fields. The potential for low-cost evaluation is substantial. Administrative records cost pennies per record, versus several hundred dollars (at least) for high-quality survey data. Administrative data also offer the potential to study population-level data. Administrative data sets are of a scale that make it possible to more precisely identify sta-

---

[3] Administrative records could include student transcripts (with student demographic data, academic performance indicators, and scholarship, graduation, and curriculum information), enrollment statistics, budgetary plans, and personnel demographic data.

> **BOX 5.1**
> **How Data Have Transformed the Study of Education**
>
> The use of data science to study education is not new. While much education research has used survey data, the increase in cost and decline in quality of surveys (Japec et al., 2015), combined with the availability of new types of data, have resulted in a transformation in education research.
>
> Examples of the large number of questions on the impacts of various differences or interventions on student outcomes (Figlio, Karbownik, and Salvanes, 2016) are many: "Registry data have been used to study the introduction of new technologies to schools in England (Machin, McNally, and Silva, 2007), experimental evidence on schools' influence on parents' involvement in education in France (Avvisati et al., 2014), experimental evidence on gender differences in competitiveness and their consequences for educational choices in the Netherlands (Buser, Niederle, and Oosterbeek, 2014), the role of school quality in Romania (Pop-Eleches and Urquiola, 2013), experimental evidence on learning incentives in Mexico (Behrman et al., 2015), perceived effects of school quality on the housing market (Figlio and Lucas, 2004), the role of peer effects utilizing student reshuffling owing to extreme events (Imberman, Kugler, and Sacerdote, 2012), or the ability of principals to recognize effective teachers (Jacob and Lefgren, 2008)" (Figlio, Karbownik, and Salvanes, 2016).
>
> One of the most compelling results of using administrative data to trace education pathways is evident in the work of Stanford University's Raj Chetty. His work looking at the role of colleges in intergenerational mobility,[1] for example, could be replicated at a smaller scale, as his class[2] demonstrates.
>
> ---
>
> [1] A list of Chetty's publications can be found at http://www.equality-of-opportunity.org/documents/, accessed February 15, 2018.
> [2] The website for Chetty's class is http://www.equality-of-opportunity.org/bigdatacourse/, accessed February 15, 2018.

tistical relationships, detect rare events, create comparison groups, and examine the heterogeneous effects of educational policies and practice for different groups of individuals. This last opportunity is particularly important when trying to encourage the broad participation of underrepresented student populations. Administrative data are also near real time in nature—it is possible to identify issues that may arise with particular interventions and adjust accordingly rather than waiting long after the fact. Last, administrative data are often of higher quality than survey data. Rather than asking respondents to provide retrospective information, with obvious recall problems, the data exist within administrative systems. It is possible, through links to state wage record, Internal Revenue Service, or Census Bureau data, to track the earnings and employment

BOX 5.2
Linking Administrative Data

Academic institutions have a wealth of administrative data that can be used for evaluation purposes with minimal cost and burden, and there is a consortium that has been established by universities to help standardize the data and make analysis possible. Administrative student transcript data, research funding data, and outcome measures have been integrated at the Institute for Research on Innovation and Science (IRIS) and through links to the U.S. Census Bureau and tax records on earnings and employment outcome. Relevant data include the following:

1. Student enrollment data, which include basic demographics and information on academic performance, are available in all academic institutions. In addition to information on the characteristics at entry of individual students in all fields, such data allow for the construction of detailed contextual information at the level of cohorts. At IRIS, for example, data currently exist for the University of California system, the University of Wisconsin, and the University of Michigan. A concerted effort could be made to incorporate additional institutions that are implementing data science programs at the undergraduate and graduate levels.
2. Research expenditures data include data on undergraduate students who participate in research. For example, UMETRICS data, which are hosted by IRIS, report information on all people paid by federal research grants and enable construction of detailed contextual measures of research training, especially laboratories and researcher networks (Lane et al., 2015). The number of universities participating in UMETRICS is increasing.[1]
3. Dissertation data, such as those derived from ProQuest,[2] provide information on the scientific fields of graduate training as well as identify shared advisors and committee members.
4. Grant titles and abstracts data, such as those extracted from the Dimensions database,[3] allow analysis of topics and classification by fields of research.
5. Census Bureau data, including the Longitudinal Employer–Household Dynamics (LEHD) data set and the Longitudinal Business Database (LBD), permit the examination of the career outcomes, such as earnings and employer characteristics, of students who complete or do not complete their degrees. The data can also be used to construct proxy measures of labor market expectations for each individual—based both on labor market conditions (from the LEHD Quarterly Workforce Indicator series) and on the earnings and employment outcomes of individuals in their peer groups and networks.

Together, such linked administrative records give unparalleled sample size and detail to study the pathways of students, including underrepresented minority and female students, in all fields.

---

[1] To view the current list, see http://iris.isr.umich.edu/, accessed January 22, 2018.
[2] The website for ProQuest is http://www.proquest.com/, accessed January 22, 2018.
[3] The website for the Dimensions database is https://www.uberresearch.com/visual-portfolios/, accessed January 22, 2018.

outcomes of individuals longitudinally with much less attrition (and cost) than surveys.

Finding 5.4: The evolution of undergraduate education in data science can be driven by data science. Exploiting administrative records, in conjunction with other data sources such as economic information and survey data, can enable effective transformation of programs to better serve their students.

Evaluation More Broadly

Unfortunately, data that would allow the understanding of outcomes from multiple programs (or even, in fact, of a single program from outside that institution) have not previously existed in a readily available form. For example, even within an institution there are regular laments about the lack of tracking data about its graduates. Accessing the pathways that undergraduates or newly graduated students in data science follow would offer potential benefit not only for an institution, relative to the evaluation criteria for its programs and institutional strategies, but also offer the possibility, when adequately shared, to evaluate broadly the impact of major national programs at the National Science Foundation and other agencies and private foundations (see Box 5.2).

Fortunately, new data have become available to permit, for the first time, the undertaking of rich analyses. The data also enable an entirely new approach to estimating the impact of labor market demand on career pathways (University of Texas System, 2016; AIR, 2018; Heckman, 2010). The advantage is that not only can these new data compare data science graduates to one another, but they can also compare the career pathways, successes, and failures of data science graduates to those of graduates in other fields. Such information can then be used to motivate further development and revision of the evolving data science curricula at the undergraduate level.

Empirical evaluation is now possible because of the integration of multiple disparate sources of data at many academic institutions—for example, the Ohio State University Ohio Education Research Center,[4] which is a collaboration of six Ohio universities and four research organizations and has data on students from preschool through the workforce. Seven states—Arkansas, Colorado, Minnesota, Tennessee, Texas, Virginia, and Washington—have linked data so that students can see the earnings of graduates from institutions within these states (see Selingo, 2017).

---

[4] The website for the Ohio Education Research Center is http://oerc.osu.edu, accessed March 29, 2018.

When integrated, these sources provide a unique data set for examining the educational, scholarly, and career trajectories of students while providing new insights into the mechanisms that shape outcomes for students from different backgrounds, races and ethnicities, and genders. These data include measures of individual characteristics ranging from undergraduate institution and standardized exam scores to race, ethnicity, and some information on financial status. From such data, it becomes possible to identify cohorts at various levels, to characterize networks that provide social support as well as access to role models, and to provide information and resources that could contribute to career success. These data can also illuminate both the labor market outcomes and shocks to demand that can drive outcomes for recent graduates. In the conceptual framework outlined earlier, students' own characteristics and backgrounds are some of the most important determinants of outcomes either on their own or in combination with their courses, internships, and other educational experiences. These new data also permit the measurement of these important characteristics of students.

Finding 5.5: Data science methods applied both to individual programs and comparatively across programs can be used for both evaluation and evolution of data science program components. It is essential that both processes are sustained as new pathways emerge at institutions.

Recommendation 5.3: Academic institutions should ensure that programs are continuously evaluated and should work together to develop professional approaches to evaluation. This should include developing and sharing measurement and evaluation frameworks, data sets, and a culture of evolution guided by high-quality evaluation. Efforts should be made to establish relationships with sector-specific professional societies to help align education evaluation with market impacts.

ROLES FOR PROFESSIONAL SOCIETIES

There are several challenges facing educational programs in any emerging field, and data science is no exception. As programs emerge and evolve, they look for best practices, class materials, guidance on career paths for students, and a network or community with which to share ideas. Professional societies have a role to play in facilitating the community building and resource sharing that are needed.

Data science benefits from sharing of materials whenever possible. However, there has been pushback to material collection by some researchers. Many may not have had intellectual property rights to these materials and did not want to share materials that were not theirs; others have had concerns about loss of intellectual property. Still, it is important to learn from others' attempts. It is also important to have a means to identify and fill gaps in a program's resources (e.g., workflow, how to link together the tools and use them in real projects). Professional societies may play a major role in making materials available and in enabling educators to receive credit for sharing and material reuse. By valuing the high-quality application of data over the quantity of data that is collected, this effort on the part of professional societies could then incentivize even more data sharing. Professional societies can also play an important role in educating people about the *process* for sharing data by publishing best practices for increasing access to and awareness of useful, high-quality data sets.

Particularly for rapidly evolving fields such as data science, access to information about career paths (whether pursuing graduate education or moving into the workforce directly) may be difficult for students to obtain, significantly harder than for students in other long-standing fields. For example, many STEM areas actively recruit undergraduates to attend professional society conferences at which they offer a wide range of career option workshops and presentations. Students in data science fields may find that obtaining such information is more difficult than for their peers in various STEM disciplines with large and active professional societies. Beyond formal meetings, professional societies host a wide array of information on their websites that informs students about career options.

The data science community would be well served by having a steering body to help organize conferences, collect materials (e.g., for programs, evaluation[5]), provide models for industry interactions, and undertake other activities. Professional societies could explore their role in pushing industry to provide career paths for students in order to motivate student adoption of data science programs. STEM-minded students by and large can see the benefit in being trained as data scientists, especially in this competitive job environment.

Retention of talent, consequently, is a real challenge in most settings. New models may be needed to advance career progression and perhaps rotational models that allow for work flexibility and expanded problem sets. A clearinghouse or professional society, perhaps with foundation support, could help to foster further connection among the variety of established programs. However, it may be too difficult for a single data

---

[5] See, for example, the American Evaluation Association (http://www.eval.org/) or the Association for Public Policy Analysis and Management (http://www.appam.org/), accessed January 22, 2018.

science professional society to represent and be relevant to all data science communities. It is important to be cognizant and supportive of the many data scientists likely to be working in other fields. In addition to the need to accommodate the multidisciplinarity of data science, there is also a need to consider the business case for establishing new societies. It is unlikely that the revenue required for such an endeavor exists. A more structured collaboration of existing professional societies that work well together might be more effective; subsocieties devoted to data science elements may develop in any of these societies. These subsocieties could be closely connected to the educational opportunities for their members, fostering a sense of community and improved professional development, while also promoting connections among practicing data scientists in other fields. Whatever convening mechanisms are chosen, they could cover varied topics such as curriculum, on-ramping, continuing education, and bringing in multiple viewpoints, such as those from industry. Other opportunities include syndication, filtering, and editorial roles.

Finding 5.6: As professional societies adapt to data science, improved coordination could offer new opportunities for additional collaboration and cross-pollination. A group or conference with bridging capabilities would be helpful. Professional societies may find it useful to collaborate to offer such training and networking opportunities to their joint communities.

Recommendation 5.4: Existing professional societies should coordinate to enable regular convening sessions on data science among their members. Peer review and discussion are essential to share ideas, best practices, and data.

REFERENCES

Aaronson, D., L. Barrow, and W. Sander. 2007. Teachers and student achievement in the Chicago public high schools. *Journal of Labor Economics* 25(1):95-135.

AIR (American Institutes for Research). 2018. "College Measures: Improving Higher Education Outcomes in the United States." http://www.air.org/center/college-measures/. Accessed April 17, 2018.

Avvisati, F., M. Guragand, N. Guyon, and E. Maurin. 2014. Getting parents involved: A field experiment in deprived schools. *Review of Economic Studies* 81(1):57-83.

Behrman, J.R., S.W. Parker, P.E. Todd, and K.I. Wolpin. 2015. Aligning learning incentives of students and teachers: Results from a social experiment in Mexican high schools. *Journal of Political Economy* 123(2):325-364.

Buser, T., M. Niederle, and H. Oosterbeek. 2014. Gender, competitiveness, and career choices. *Quarterly Journal of Economics* 129(3):1409-1447.

Figlio, D., K. Karbownik, and K.G. Salvanes. 2016. Education research and administrative data. *Handbook of the Economics of Education* 5:75-138.

Figlio, D.N., and M.E. Lucas. 2004. Do high grading standards affect student performance? *Journal of Public Economics* 89:1815-1834.

Gertler, P.J., S. Martinez, P. Premand, L.B. Rawlings, and C.M.J. Vermeersch. 2016. *Impact Evaluation in Practice*. 2nd ed. Washington, D.C.: Inter-American Development Bank and World Bank.

Heckman, J.J. 2010. Building bridges between structural and program evaluation approaches to evaluating policy. *Journal of Economic Literature* 48(2):356-398.

Imberman, S.A., A.D. Kugler, and B.I. Sacerdote. 2012. Katrina's children: Evidence on the structure of peer effects from hurricane evacuees. *American Economic Review* 102(5):2048-2082.

Jacob, B.A., and L. Lefgren. 2008. Can principals identify effective teachers? Evidence on subjective performance evaluation in education. *Journal of Labor Economics* 26(1):101-136.

Japec, L., F. Kreuter, M. Berg, P. Biemer, P. Decker, C. Lampe, J. Lane, C. O'Neil, and A. Usher. 2015. Big data in survey research: AAPOR Task Force report. *Public Opinion Quarterly* 79(4):839-880.

Lane, J., J. Owen-Smith, R. Rosen, and B. Weinberg. 2015. New linked data on research investments: Scientific workforce, productivity, and public value. *Research Policy* 44(9):1659-1671.

Machin, S., S. McNally, and O. Silva. 2007. New technology in schools: Is there a payoff? *Economic Journal* 117(522):1145-1167.

Metcalf, H. 2010. Stuck in the pipeline: A critical review of STEM workforce literature. *InterActions: UCLA Journal of Education and Information Studies* 6(2):1-20.

National Science Board. 2015. *Revisiting the STEM Workforce*. https://www.nsf.gov/nsb/publications/2015/nsb201510.pdf. Accessed January 23, 2018.

NRC (National Research Council). 2012. *Discipline-Based Education Research: Understanding and Improving Learning in Undergraduate Science and Engineering*. Washington, D.C.: The National Academies Press.

NSF (National Science Foundation). 2014. "College Board Launches New AP Computer Science Principles Course." https://www.nsf.gov/news/news_summ.jsp?cntn_id=133571. Accessed February 13, 2018.

Pop-Eleches, C., and M. Urquiola. 2013. Going to a better school: Effects and behavioral responses. *American Economic Review* 103(4):1289-1324.

Selingo, J.J. 2017. Six myths about choosing a college major. *New York Times*, November 3. https://nyti.ms/2iYZN3r. Accessed January 22, 2018.

UC Berkeley (University of California, Berkeley). 2018. "The BJC Curriculum." https://bjc.berkeley.edu/curriculum/. Accessed February 13, 2018.

University of Texas System. 2016. UT System partners with U.S. Census Bureau to provide salary and jobs data of UT graduates across the nation. Press release, September 22. https://www.utsystem.edu/news/2016/09/22/ut-system-partners-us-census-bureau-provide-salary-and-jobs-data-ut-graduates-across. Accessed February 13, 2018.

6

Conclusions

Data science education is well into its formative stages of development; it is evolving into a self-supporting discipline and producing professionals with distinct and complementary skills relative to professionals in the computer, information, and statistical sciences. However, regardless of its potential eventual disciplinary status, the evidence points to robust growth of data science education that will indelibly shape the undergraduate students of the future. In fact, fueled by growing student interest and industry demand, data science education will likely become a staple of the undergraduate experience. There will be an increase in the number of students majoring, minoring, earning certificates, or just taking courses in data science as the value of data skills becomes even more widely recognized. The adoption of a general education requirement in data science for all undergraduates will endow future generations of students with the basic understanding of data science that they need to become responsible citizens. Continuing education programs such as data science boot camps, career accelerators, summer schools, and incubators will provide another stream of talent. This constitutes the emerging watershed of data science education that feeds multiple streams of generalists and specialists in society; citizens are empowered by their basic skills to examine, interpret, and draw value from data.

Today, the nation is in the formative phase of data science education, where educational organizations are pioneering their own programs, each with different approaches to depth, breadth, and curricular emphasis (e.g., business, computer science, engineering, information science, mathematics, social science, or statistics). It is too early to expect consensus to emerge on certain best practices of data science education. However, it is not too early to envision the possible forms that such practices might take. Nor is it too early to make recommendations that can help the data science education community develop strategic vision and practices. The following is a summary of the findings and recommendations discussed in the preceding four chapters of this report.

Finding 2.1: Data scientists today draw largely from extensions of the "analyst" of years past trained in traditional disciplines. As data science becomes an integral part of many industries and enriches research and development, there will be an increased demand for more holistic and more nuanced data science roles.

Finding 2.2: Data science programs that strive to meet the needs of their students will likely evolve to emphasize certain skills and capabilities. This will result in programs that prepare different types of data scientists.

Recommendation 2.1: Academic institutions should embrace data science as a vital new field that requires specifically tailored instruction delivered through majors and minors in data science as well as the development of a cadre of faculty equipped to teach in this new field.

Recommendation 2.2: Academic institutions should provide and evolve a range of educational pathways to prepare students for an array of data science roles in the workplace.

Finding 2.3: A critical task in the education of future data scientists is to instill data acumen. This requires exposure to key concepts in data science, real-world data and problems that can reinforce the limitations of tools, and ethical considerations that permeate many applications. Key concepts involved in developing data acumen include the following:

- Mathematical foundations,
- Computational foundations,
- Statistical foundations,
- Data management and curation,
- Data description and visualization,
- Data modeling and assessment,
- Workflow and reproducibility,

- Communication and teamwork,
- Domain-specific considerations, and
- Ethical problem solving.

Recommendation 2.3: To prepare their graduates for this new data-driven era, academic institutions should encourage the development of a basic understanding of data science in all undergraduates.

Recommendation 2.4: Ethics is a topic that, given the nature of data science, students should learn and practice throughout their education. Academic institutions should ensure that ethics is woven into the data science curriculum from the beginning and throughout.

Recommendation 2.5: The data science community should adopt a code of ethics; such a code should be affirmed by members of professional societies, included in professional development programs and curricula, and conveyed through educational programs. The code should be reevaluated often in light of new developments.

Finding 3.1: Undergraduate education in data science can be experienced in many forms. These include the following:

- Integrated introductory courses that can satisfy a general education requirement;
- A major in data science, including advanced skills, as the primary field of study;
- A minor or track in data science, where intermediate skills are connected to the major field of study;
- Two-year degrees and certificates;
- Other certificates, often requiring fewer courses than a major but more than a minor;
- Massive open online courses, which can engage large numbers of students at a variety of levels; and
- Summer programs and boot camps, which can serve to supplement academic or on-the-job training.

Recommendation 3.1: Four-year and two-year institutions should establish a forum for dialogue across institutions on all aspects of data science education, training, and workforce development.

Finding 4.1: The nature of data science is such that it offers multiple pathways for students of different backgrounds to engage at levels ranging from basic to expert.

Finding 4.2: Data science would particularly benefit from broad participation by underrepresented minorities because of the many applications to problems of interest to diverse populations.

Recommendation 4.1: As data science programs develop, they should focus on attracting students with varied backgrounds and degrees of preparation and preparing them for success in a variety of careers.

Finding 4.3: Institutional flexibility will involve the development of curricula that take advantage of current course availability and will potentially be constrained by the availability of teaching expertise. Whatever organizational or infrastructure model is adopted, incentives are needed to encourage faculty participation and to overcome barriers.

Finding 4.4: The economics of developing programs has recently changed with the shift to cloud-based approaches and platforms.

Finding 5.1: The evolution of data science programs at a particular institution will depend on the particular institution's pedagogical style and the students' backgrounds and goals, as well as the requirements of the job market and graduate schools.

Recommendation 5.1: Because these are early days for undergraduate data science education, academic institutions should be prepared to evolve programs over time. They should create and maintain the flexibility and incentives to facilitate the sharing of courses, materials, and faculty among departments and programs.

Finding 5.2: There is a need for broadening the perspective of faculty who are trained in particular areas of data science to be knowledgeable of the breadth of approaches to data science so that they can more effectively educate students at all levels.

Recommendation 5.2: During the development of data science programs, institutions should provide support so that the faculty can become more cognizant of the varied aspects of data science through discussion, co-teaching, sharing of materials, short courses, and other forms of training.

Finding 5.3: The data science community would benefit from the creation of websites and journals that document and make available best practices, curricula, education research findings, and other materials related to undergraduate data science education.

Finding 5.4: The evolution of undergraduate education in data science can be driven by data science. Exploiting administrative records, in conjunction with other data sources such as economic information and survey data, can enable effective transformation of programs to better serve their students.

Finding 5.5: Data science methods applied both to individual programs and comparatively across programs can be used for both evaluation and evolution of data science program components. It is essential that both processes are sustained as new pathways emerge at institutions.

Recommendation 5.3: Academic institutions should ensure that programs are continuously evaluated and should work together to develop professional approaches to evaluation. This should include developing and sharing measurement and evaluation frameworks, data sets, and a culture of evolution guided by high-quality evaluation. Efforts should be made to establish relationships with sector-specific professional societies to help align education evaluation with market impacts.

Finding 5.6: As professional societies adapt to data science, improved coordination could offer new opportunities for additional collaboration and cross-pollination. A group or conference with bridging capabilities would be helpful. Professional societies may find it useful to collaborate to offer such training and networking opportunities to their joint communities.

Recommendation 5.4: Existing professional societies should coordinate to enable regular convening sessions on data science among their members. Peer review and discussion are essential to share ideas, best practices, and data.

Appendixes

A

Biographies of the Committee

LAURA HAAS, *Co-Chair*, joined the University of Massachusetts Amherst in August 2017 as dean of the College of Information and Computer Sciences, after a long career at IBM, where she was accorded the title IBM Fellow in recognition of her impact. At the time of Dr. Haas's retirement from IBM, she was director of IBM Research's Accelerated Discovery Lab (2011-2017), after serving as director of computer science at IBM's Almaden Research Center from 2005 to 2011. She had worldwide responsibility for IBM Research's exploratory science program from 2009 through 2013. From 2001 to 2005, she led the Information Integration Solutions architecture and development teams in IBM's Software Group. Previously, Dr. Haas was a research staff member and manager at Almaden. She is best known for her work on the Starburst query processor, from which the database server DB2 LUW was developed, on Garlic, a system that allowed integration of heterogeneous data sources, and on Clio, the first semiautomatic tool for heterogeneous schema mapping. She has received several IBM awards for Outstanding Innovation and Technical Achievement, an IBM Corporate Award for Information Integration Technology, the Anita Borg Institute Technical Leadership Award, and the Association for Computing Machinery (ACM) Special Interest Group on Management of Data Edgar F. Codd Innovation Award. Dr. Haas was vice president of the Very Large Data Bases Endowment Board of Trustees from 2004 to 2009 and served on the board of the Computing Research Association from 2007 to 2016 (vice chair, 2009-2015). She currently serves on the National Academies Computer Science and Telecommunications Board

(2013- 2019). She is an ACM fellow, a member of the National Academy of Engineering, and a fellow of the American Academy of Arts and Sciences.

ALFRED O. HERO III, *Co-Chair*, is the John H. Holland Distinguished University Professor of Electrical Engineering and Computer Science and R. Jamison and Betty Williams Professor of Engineering at the University of Michigan. Dr. Hero's primary appointment is in the Department of Electrical Engineering and Computer Science, and he also has appointments, by courtesy, in the Department of Biomedical Engineering and the Department of Statistics. In 2008, Dr. Hero was awarded the Digiteo Chaire d'Excellence, sponsored by Digiteo Research Park in Paris, located at the Ecole Supérieure d'Electricité, Gif-sur-Yvette, France. He is an Institute of Electrical and Electronics Engineers (IEEE) fellow, and several of his research articles have received best paper awards. Dr. Hero was awarded the University of Michigan Distinguished Faculty Achievement Award (2011). He received the IEEE Signal Processing Society Meritorious Service Award (1998) and the IEEE Third Millennium Medal (2000). He was president of the IEEE Signal Processing Society (2006-2008) and was on the board of directors of the IEEE (2009-2011), where he served as director of Division IX (Signals and Applications). Dr. Hero's recent research interests have been in detection, classification, pattern analysis, and adaptive sampling for spatiotemporal data. Of particular interest are applications to network security, multimodal sensing and tracking, biomedical imaging, and genomic signal processing. Dr. Hero received a B.S. (summa cum laude) from Boston University in 1980 and a Ph.D. from Princeton University in 1984, both in electrical engineering.

ANI ADHIKARI is a senior lecturer in statistics at the University of California, Berkeley. Dr. Adhikari has received the Distinguished Teaching Award at Berkeley and the Dean's Award for Distinguished Teaching at Stanford University. While Dr. Adhikari's research interests are centered on applications of statistics in the natural sciences, her primary focus has always been on teaching and mentoring students. She teaches courses at all levels and has a particular affinity for teaching statistics to students who have little mathematical preparation. Dr. Adhikari received an undergraduate degree from the Indian Statistical Institute and a Ph.D. in statistics from the University of California, Berkeley.

DAVID CULLER joined the faculty of the University of California, Berkeley, in 1989 and is currently the interim dean for Data Sciences. He was associate chair (2010-2012) and chair (2012-June 2014) of the Electrical Engineering and Computer Sciences department, the founding director of Intel Research, Berkeley, and co-founder of Arch Rock Corp.

Dr. Culler won the Okawa Prize in 2013. He is a member of the National Academy of Engineering, an Association for Computing Machinery fellow, and an Institute of Electrical and Electronics Engineers fellow. He has been named one of *Scientific American*'s "Top 50 Researchers" and the creator of one of *MIT Technology Review*'s "10 Technologies That Will Change the World." Dr. Culler's research addresses distributed systems for the built environment, networks of small, embedded wireless devices; planetary-scale Internet services; parallel computer architecture; parallel programming languages; and high-performance communication. Dr. Culler received a B.A. from the University of California, Berkeley, in 1980, and an M.S. and a Ph.D. from the Massachusetts Institute of Technology in 1985 and 1989, respectively.

DAVID DONOHO is an Anne T. and Robert M. Bass Professor of Humanities and Sciences and a professor of statistics at Stanford University. He is a member of the National Academy of Sciences and Foreign Associate of the French Académie des Sciences. He has worked in industrial research in oil exploration (Western Geophysical), in empirical finance (Renaissance Technologies), and co-founded a successful information technology start-up, BigFix, acquired by IBM. Dr. Donoho received an A.B. from Princeton University and a Ph.D. from Harvard University. He received a Macarthur Fellowship and was recipient of the 2013 Shaw Prize in Mathematical Sciences.

E. THOMAS EWING is an associate dean for Graduate Studies, Research, and Diversity in the College of Liberal Arts and Human Sciences and a professor in the Department of History at Virginia Tech. Dr. Ewing teaches courses in the history of data in social context, as well as courses in Russian, European, Middle Eastern, and world history; gender/women's history; and historical methods. His publications include *Separate Schools: Gender, Policy, and Practice in the Postwar Soviet Union* (2010) and *The Teachers of Stalinism: Policy, Practice, and Power in Soviet Schools in the 1930s* (2002), and his articles on Stalinist education have been published in *Gender & History*, *American Educational Research Journal*, *Women's History Review*, *History of Education Quarterly*, *Russian Review*, and *Journal of Women's History*. He has received six grants from the National Endowment for the Humanities at the intersection of digital humanities, data analytics, and medical history projects. Dr. Ewing's education includes a B.A. from Williams College and a Ph.D. in history from the University of Michigan.

LOUIS J. GROSS is an Alvin and Sally Beaman Distinguished Professor of Ecology and Evolutionary Biology and Mathematics and director of the Institute for Environmental Modeling at the University of Tennessee,

Knoxville (UTK). Dr. Gross is also director of the National Institute for Mathematical and Biological Synthesis, a National Science Foundation-funded center to foster research and education at the interface between math and biology. His research focuses on applications of mathematics and computational methods in many areas of ecology, including disease ecology, landscape ecology, spatial control for natural resource management, photosynthetic dynamics, and the development of quantitative curricula for life science undergraduates. Dr. Gross led the effort at UTK to develop an across-trophic-level modeling framework to assess the biotic impacts of alternative water planning for the Everglades of Florida. He has co-directed several courses and workshops in mathematical ecology at the International Centre for Theoretical Physics in Trieste, Italy, and he has served as program chair of the Ecological Society of America, president of the Society for Mathematical Biology, president of the UTK Faculty Senate, treasurer for the American Institute of Biological Sciences, and chair of the National Research Council Committee on Education in Biocomplexity Research. Dr. Gross is the 2006 Distinguished Scientist awardee of the American Institute of Biological Sciences and is a fellow of the American Association for the Advancement of Science and of the Society for Mathematical Biology. He has served on the National Research Council Board on Life Sciences and was liaison to the National Research Council Standing Committee on Emerging Science for Environmental Health Decisions. Dr. Gross completed a B.S. in mathematics at Drexel University and a Ph.D. in applied mathematics at Cornell University, and has been a faculty member at UTK since 1979.

NICHOLAS J. HORTON is Beitzel Professor of Technology and Society and professor of statistics at Amherst College. Dr. Horton is an applied biostatistician whose work is based squarely within the mathematical and computational sciences but spans other fields in order to ensure that biomedical research is conducted on a sound footing. He has published more than 170 papers in the statistics and biomedical literature and 4 books on statistical computing and data science. He has taught a variety of courses in statistics and related fields, including introductory statistics, data science, probability, theoretical statistics, regression, and design of experiments. He is passionate about improving quantitative and computational literacy for students with a variety of backgrounds as well as engagement and mastery of higher level concepts and capacities to think with data. Dr. Horton received the American Statistical Association (ASA) Waller Award for Distinguished Teaching, the Mathematical Association of America Hogg Award for Excellence in Teaching, the Mu Sigma Rho Statistics Education Award, and the ASA Founders Award. He was a co-principal investigator of the National Science Foundation-funded Project MOSAIC, serves as the chair of the Committee of Presidents of Statistical Societies, is a fellow of the ASA and the American Association for the Advancement of Science, and was a research fellow at the Bureau of Labor Statistics. Dr. Horton earned an A.B. from Harvard College and an Sc.D. in biostatistics from the Harvard T.H. Chan School of Public Health.

JULIA LANE is a professor at the Center for Urban Science and Progress (CUSP) and at the New York University (NYU) Wagner Graduate School of Public Service. Dr. Lane also serves as a provostial fellow for innovation analytics and senior fellow at NYU's GovLab. Dr. Lane is an economist who is the co-founder of the Longitudinal Employer–Household Dynamics (LEHD) partnership with the Census Bureau, which is now a major national statistical program. Dr. Lane has authored almost 80 refereed articles and edited or authored 10 books. Dr. Lane's work to quantify the results of federal stimulus spending has been published in *Science* and *Nature*. She co-founded the new Institute for Research on Innovation and Science at the University of Michigan, which uses empirical evidence to describe the structure of the research workforce, the nature and evolution of research collaborations, and the diffusion of sponsored research results. Dr. Lane has had leadership positions in a number of policy and data science initiatives at her other previous appointments, which include senior fellow and senior managing economist at the American Institutes for Research; senior vice president and director, Economics Department at NORC/University of Chicago; various consultancy roles at the World Bank; and assistant, associate, and full professor at American University. Dr. Lane received a Ph.D. in economics and a master's in statistics from the University of Missouri.

ANDREW McCALLUM is a professor and director of the Center for Data Science, as well as the Information Extraction and Synthesis Laboratory, in the College of Information and Computer Science at the University of Massachusetts Amherst. Dr. McCallum has published over 250 papers in many areas of artificial intelligence, including natural language processing, machine learning, and reinforcement learning; his work has received over 45,000 citations. In the early 2000s he was vice president of research and development at WhizBang Labs, a 170-person start-up company that used machine learning for information extraction from the web. Dr. McCallum is an Association for the Advancement of Artificial Intelligence fellow and the recipient of the UMass Chancellor's Award for Research and Creative Activity, the UMass NSM Distinguished Research Award, the UMass Lilly Teaching Fellowship, as well as research awards from Google, IBM, Microsoft, and Yahoo. He was the general chair for the International Conference on Machine Learning 2012 and is the current president of the International Machine Learning Society, as well as member of the editorial board of the *Journal of Machine Learning Research*. For the past 10 years, Dr. McCallum has been active in research on statistical machine learning applied to text, especially information extraction, entity resolution, social network analysis, structured prediction, semisupervised learning, and deep neural networks for knowledge representation. Dr. McCallum obtained a Ph.D. from the University of Rochester in 1995 with Dana Ballard and a postdoctoral fellowship from Carnegie Mellon University with Tom Mitchell and Sebastian Thrun.

RICHARD McCULLOUGH has been the vice provost for research (VPR) at Harvard University since 2012, working with the president and provost to encourage, cultivate, and coordinate high-impact academic research across all of Harvard's schools and affiliated institutions. The office of the VPR has broad responsibility and oversight for the development, review, and implementation of strategies, planning, and policies related to the organization and execution of academic research across the entire university. Dr. McCullough leads a new office of foundation and corporate development. He also assists in oversight of many of the interdisciplinary institutes, centers, and initiatives across Harvard. Under Dr. McCullough's leadership, the office of the VPR is particularly focused on removing barriers to collaboration, whether in university policies or in financial or administrative systems. Additionally, the VPR works with the president and provost to foster and encourage entrepreneurship among undergraduates, graduate students, and faculty members, and helps to lead the development of the new innovation campus. Dr. McCullough is also a professor of materials science and engineering at Harvard and is a member of numerous professional societies and boards. Prior to being named VPR at Harvard, Dr. McCullough was the vice president for research at Carnegie Mellon University in Pittsburgh, where he previously served as the dean of the Mellon College of Science and professor and head of the Department of Chemistry. Dr. McCullough has founded two companies: Plextronics, Inc., and Liquid X Printed Metals. Dr. McCullough has a B.S. in chemistry from the University of Texas, Dallas, and earned an M.A. and a Ph.D. in chemistry at Johns Hopkins University. He did his postdoctoral fellowship at Columbia University.

REBECCA NUGENT is the associate department head, the director of undergraduate studies, and a teaching professor in the Department of Statistics & Data Science at Carnegie Mellon University (CMU) and has been teaching at CMU since 2006. Dr. Nugent recently was awarded top teaching honors with the American Statistical Association Waller Education Award; the William H. and Frances S. Ryan Award for Meritorious Teaching; and Statistician of the Year by the ASA Pittsburgh Chapter. Dr. Nugent's research interests lie in clustering, record linkage, educational data mining/psychometrics, public health, tech/innovation/entrepreneurship, and semantic organization. Dr. Nugent received a B.A with majors in mathematics, statistics, and Spanish at Rice University and an M.S. in statistics at Stanford University. She completed a Ph.D. in statistics at the University of Washington in 2006.

LEE RAINIE is the director of Internet, science, and technology research at Pew Research Center. Under Mr. Rainie's leadership, the center has issued more than 500 reports based on its surveys that examine people's online activities and the role of the Internet in their lives. Mr. Rainie also directs the center's new initiative on the intersection of science and society. The American Sociological Association gave Mr. Rainie its award for Excellence in Reporting on Social Issues in 2014 and described his work as the "most authoritative source of reliable data on the use and impact of the internet and mobile connectivity." Rainie is a co-author of *Networked: The New Social Operating System* (MIT Press, 2012) and five books about the future of the Internet that are drawn from the center's research. Mr. Rainie gives several dozen speeches a year to government officials, media leaders, scholars and students, technology executives, librarians, and non-profit groups about the changing media ecosystem. He is also regularly interviewed by major news organizations about technology trends. Prior to launching Pew Research Center's technology research, Mr. Rainie was managing editor of *U.S. News & World Report*. He is a graduate of Harvard University and has a master's degree in political science from Long Island University.

ROB RUTENBAR joined the faculty at Carnegie Mellon University (CMU) in 1985. Dr. Rutenbar spent 25 years in electrical and computer engineering at CMU, ultimately holding the Stephen J. Jatras (E'47) Chair. He was the founding director of the Center for Circuit and System Solutions, a large consortium of U.S. schools (e.g., CMU, Massachusetts Institute of Technology, Stanford, Berkeley, Caltech, Cornell, Columbia, Georgia Tech, University of California, Los Angeles) supported by the Defense Advanced Research Projects Agency and the U.S. semiconductor industry, focused on design problems at the end of Moore's Law scaling. In 2010, Dr. Rutenbar moved to the University of Illinois, Urbana-Champaign, where he was Abel Bliss Professor and head of the Department of Computer Science. At the University of Illinois, he pioneered the novel CS + X program, which combines a core computer science curriculum with a disciplinary "X" curriculum, leading to a bachelor's degree in "X." Student pipelines for CS + anthropology, astronomy, chemistry, and linguistics are now under way, with several more CS + X degrees under design. Dr. Rutenbar now serves as senior vice chancellor for research at the University of Pittsburgh. His research has focused on three primary areas: tools for integrated circuit design, statistics of nanoscale chip designs, and custom architectures for machine learning and perception. In 1998 he founded Neolinear, Inc., to commercialize the first practical synthesis tools for nondigital ICs, and served as Neolinear's chief scientist until its acquisition by Cadence in 2004. In 2006, he founded Voci Technologies, Inc., to commercialize enterprise-scale voice analytics. Dr. Rutenbar has won numerous awards, including the Institute of Electrical and Electronics Engineers Circuits and Systems Society Industrial Pioneer Award and the Semiconductor Research Corporation Aristotle Award. His work has been featured in venues ranging from *Slashdot* to the *Economist*. Dr. Rutenbar received a Ph.D. from the University of Michigan in 1984.

KRISTIN M. TOLLE is the director of the Data Science Initiative in Microsoft Research Outreach, Redmond, Washington. Dr. Tolle joined Microsoft in 2000 and has acquired numerous patents and worked for several product teams including the Natural Language Group, Visual Studio, and the Microsoft Office Excel Team. Since joining the Microsoft Research Outreach program in 2006, she has run several major initiatives from biomedical computing and environmental science to more traditional computer and information science programs around natural user interactions and data curation. She also directed the development of the Microsoft Translator Hub and the Environmental Science Services Toolkit. Dr. Tolle is an editor, along with Tony Hey and Stewart Tansley, of one of the earliest books on data science, *The Fourth Paradigm: Data Intensive Scientific Discovery* (Microsoft Research, 2009). Her current focus is developing an outreach program to engage with academics on data science in general and more specifically around using data to create meaningful and useful user experiences across device platforms. Dr. Tolle's present research interests include global public health as related to climate change, mobile computing to enable field scientists and inform the public, sensors used to gather ecological and environmental data, and integration and interoperability of large heterogeneous environmental data sources. She collaborates with several major research groups in Microsoft Research including eScience, computational science laboratory, computational ecology and environmental science, and the sensing and energy research group. Prior to joining Microsoft, Dr. Tolle was an Oak Ridge Science and Engineering Research fellow for the National Library of Medicine and a research associate at the University of Arizona Artificial Intelligence Lab managing the group on medical information retrieval and natural language processing.

Dr. Tolle earned a Ph.D. in management of information systems with a minor in computational linguistics from the University of Arizona.

TALITHIA WILLIAMS takes sophisticated numerical concepts and makes them understandable and relatable to everyone. As illustrated in her popular TED Talk "Own Your Body's Data," Dr. Williams demystifies the mathematical process in amusing and insightful ways, using statistics as a way of seeing the world in a new light and transforming our future through the bold new possibilities inherent in the science, technology, engineering, and mathematics (STEM) fields. As an associate professor of mathematics at Harvey Mudd College, Dr. Williams has made it her life's work to get people—students, parents, educators, and community members—more excited about the possibilities inherent in a STEM education. In her present capacity as a faculty member, she exemplifies the role of teacher and scholar through outstanding research, with a passion for integrating and motivating the educational process with real-world statistical applications. Her professional experiences include research appointments at the Jet Propulsion Laboratory, the National Security Agency, and NASA. Dr. Williams develops statistical models that emphasize the spatial and temporal structure of data and has partnered with the World Health Organization in developing a cataract model used to predict the cataract surgical rate for countries in Africa. Through her research and work in the community at large, she is helping change the collective mind-set regarding STEM in general and math in particular—rebranding the field of mathematics as anything but dry, technical, or male dominated but instead a logical, productive career path that is crucial to the future of the country. Dr. Williams's educational background includes a bachelor's degree in mathematics from Spelman College, master's degrees both in mathematics from Howard University and in statistics from Rice University, and a Ph.D. in statistics from Rice University.

ANDREW ZIEFFLER is a senior lecturer and researcher in the Quantitative Methods in Education program within the Department of Educational Psychology at the University of Minnesota. Dr. Zieffler teaches undergraduate- and graduate-level courses in statistics and trains and supervises graduate students in statistics education. His scholarship primarily focuses on statistics education, and he has authored or co-authored several papers and book chapters related to statistics education. Additionally, he has been a co-principal investigator on many National Science Foundation-funded statistics education research projects. Dr. Zieffler has co-authored two textbooks that serve as an introduction to modern statistical and computational methods for students in the educational and behavioral sciences. He currently serves as co-editor of the journal *Technology Innovations in Statistics Education* and as a member of the Research Advisory Board for the Consortium for the Advancement of Undergraduate Statistics Education. Dr. Zieffler received his Ph.D. in quantitative methods in education from the University of Minnesota in 2006.

B

Meetings and Presentations

FIRST COMMITTEE MEETING
Washington, D.C.
December 12-13, 2016

Lessons from Current Data Science Programs and Future Directions
Rebecca Nugent, Carnegie Mellon University
Rob Rutenbar, University of Illinois, Urbana-Champaign
David Culler, University of California, Berkeley
William Yslas Velez, University of Arizona
Duncan Temple Lang, University of California, Davis

Envisioning the Field of Data Science and Future Directions and Implications to Society
David Donoho, Stanford University
Lee Rainie, Pew Research Center

Expanding Diversity in Data Science—Among Student Populations and in Topic Areas Embraced by Data Science
Bhramar Mukherjee, University of Michigan
Deb Agarwal, Lawrence Berkeley National Laboratory
Andrew Zieffler, University of Minnesota

Questions That Should Be Asked to Envision the Future of Data Science for Undergraduates
    Tom Ewing, Virginia Tech
    Louis Gross, University of Tennessee, Knoxville
    Chris Mentzel, Gordon and Betty Moore Foundation
    Patrick Perry, New York University
    John Abowd, U.S. Census Bureau

WEBINAR
April 25, 2017

Overview of the Study
    Michelle Schwalbe, National Academies of Sciences, Engineering, and Medicine
    Alfred Hero, University of Michigan
    Laura Haas, IBM Almaden Research Center
    Louis Gross, University of Tennessee, Knoxville

Facilitated Discussion
    Andy Burnett, Knowinnovation

WORKSHOP
Washington, D.C.
May 2-3, 2017

Opening Comments
    Study Co-Chairs: Laura Haas, IBM, and Alfred Hero III, University of Michigan

Comments from the National Science Foundation
    Chaitan Baru, National Science Foundation

Overview of the Workshop
    Andy Burnett, Knowinnovation

Workshop Themes

Skills and Knowledge for Future Data Scientists
    Rob Rutenbar, University of Illinois, Urbana-Champaign

Broadening Participation in Data Science Education
    Julia Lane, New York University

Future Delivery of Data Science Education
Nicholas Horton, Amherst College

Table Discussions About Key Questions

Question Exploration Groups
Small breakout groups to discuss all three questions

Feedback from Question Groups
Present ideas and discuss questions with full group

Integrate Ideas into Three Thematic Areas
Form three groups aligned with the thematic questions or possible new questions

Feedback from Question Groups
Share the integrated ideas with the full group

Plenary Discussion of Feedback
Study Co-Chairs: Laura Haas, IBM, and Alfred Hero III, University of Michigan

New Questions and Ideas That Emerged Overnight
Full group discussion led by Andy Burnett, Knowinnovation

Identify the Most Promising Ideas and Possible Findings for the Committee's Interim Report
Small table groups

Backcast the Most Promising Ideas
Small table groups discuss what steps would have to be taken in order to implement the most promising ideas

WEBINAR
BUILDING DATA ACUMEN
September 12, 2017

Capstone Courses
Nicole Lazar, University of Georgia

NC State University Data Initiative
Mladen Vouk, North Carolina State University

Moderated Discussion
Tom Ewing, Virginia Tech

WEBINAR
INCORPORATING REAL-WORLD EXAMPLES
September 19, 2017

Using Urban and Sports Data in Student Projects
Cláudio Silva, New York University

Building a Talent Pipeline Through a Strategic Career Development Program and Academic-Industrial Partnership
Sears Merritt, MassMutual Financial Group

Moderated Discussion
Tom Ewing, Virginia Tech

WEBINAR
FACULTY TRAINING AND CURRICULUM DEVELOPMENT
September 26, 2017

Go to the People: Impactful Faculty Training in Data Science
Michael Posner, Villanova University

Shodor, National Computational Science Institute, XSEDE, and Blue Waters—How Can We Help?
Bob Panoff, Shodor Education Foundation

Moderated Discussion
Nicholas Horton, Amherst College

WEBINAR
COMMUNICATION SKILLS AND TEAMWORK
October 3, 2017

The Imperative of Interdisciplinarity in Data Science
Madeleine Clare Elish, Data and Society Research Institute

Data Science Collaboration for Public-Facing Research
Adam Hughes, Pew Research Center

Moderated Discussion
Lee Rainie, Pew Research

WEBINAR
INTERDEPARTMENTAL COLLABORATION AND INSTITUTIONAL ORGANIZATION
October 10, 2017

Forging Virginia Tech's Computational Modeling and Data Analytics (CMDA) Major Across Departments
Mark Embree, Virginia Tech

Some Thoughts on Data Science Education for Undergraduates
Mike Franklin, University of Chicago

Moderated discussion
Tom Ewing, Virginia Tech

WEBINAR
ETHICS
October 17, 2017

An Ethical Reasoning Framework for Data Science Education
Sorin Adam Matei, Purdue University

Ethical Thinking for Data Science Education
Brittany Fiore-Gartland, University of Washington

Moderated Discussion
Lee Rainie, Pew Research

WEBINAR
ASSESSMENT AND EVALUATION FOR DATA SCIENCE PROGRAMS
October 24, 2017

Evaluation of Data Science Programs
Pamela Bishop, University of Tennessee, Knoxville

Assessing Data Science Learning Outcomes
Kari Jordan, Data Carpentry

Moderated Discussion
Louis Gross, University of Tennessee, Knoxville

WEBINAR
DIVERSITY, INCLUSION, AND INCREASING PARTICIPATION
November 7, 2017

Diversity, Inclusion, and Increasing Participation in Data Science
  Allison Master, University of Washington

Diversity and Inclusion in Data Science: Using Data-Informed Decisions to Drive Student Success
  Talithia Williams, Harvey Mudd College

Moderated Discussion
  Nicholas Horton, Amherst College

WEBINAR
2-YEAR COLLEGES AND INSTITUTIONAL PARTNERSHIPS
November 14, 2017

Developing a 2-year College Certificate Program in Data Science
  Brian Kotz, Montgomery College

Data Analytics Certificate Program at JCCC
  Suzanne Smith, Johnson County Community College

Moderated Discussion
  Laura Haas, University of Massachusetts Amherst

SECOND COMMITTEE MEETING
Washington, D.C.
December 6-7, 2017

Webinar Recaps
  Tom Ewing, Virginia Tech
  Nicholas Horton, Amherst College
  Lee Rainie, Pew Research
  Louis Gross, University of Tennessee, Knoxville
  Laura Haas, University of Massachusetts Amherst

Big Data Hubs
  Melissa Cragin, Midwest Big Data Hub
  Renata Rawlings-Goss, South Big Data Hub

Comments from the National Science Foundation
  Stephanie August, National Science Foundation
  Chaitan Baru, National Science Foundation

C

Contributing Individuals

The committee would also like to thank the following individuals for providing input to this study:

John Abowd, U.S. Census Bureau
Deb Agarwal, Lawrence Berkeley National Laboratory
Jon Ahlquist, Florida State University
Barbara Alvin, Eastern Washington University
Barbara Anthony, Southwestern University
Stephanie August, National Science Foundation
David Austin, North Carolina State University
Maria Aysa-Lastra, Winthrop University
Tom Barr, American Mathematical Society
Laura Bartley, University of Oklahoma
Chaitan Baru, National Science Foundation
Nina Bijedic, University "Džemal Bijedić" of Mostar
Pamela Bishop, University of Tennessee, Knoxville
Sally Blake, Flagler College
Roselie Bright, U.S. Food and Drug Administration
Quincy Brown, American Association for the Advancement of Science
Andy Burnett, Knowinnovation
Dave Campbell, Simon Fraser University
Robert Campbell, Brown University
Eva Campo, National Science Foundation
Robert Carver, Stonehill College

Amy Chang, American Society for Microbiology
Lei Cheng, Olivet Nazarene University
Hongmei Chi, Florida A&M University
Alok Choudhary, Northwestern University
William Coberly, University of Tulsa
Peyton Cook, University of Tulsa
Bill Corey, University of Virginia
Melissa Cragin, Midwest Big Data Hub
Catherine Cramer, New York Hall of Science
James Curry, University of Colorado Boulder
Nicole Dalzell, Duke University
Juliana DeCastro, Núcleo de Planejamento Estratégico de Transporte e Turismo
Sam Donovan, University of Pittsburgh
Renee Dopplick, Association for Computing Machinery
Maureen Doyle, Northern Kentucky University
Ruth Duerr, Ronin Institute
Arturo Duran, IVA Ventures
Stephen Edwards, ACM Administrative Centre
Madeleine Clare Elish, Data and Society
Sandra Ellis, Texas A&M University, Corpus Christi
Mark Embree, Virginia Polytechnic Institute and State University
Paula Faulkner, North Carolina Agricultural and Technical State University
Raya Feldman, University of California, Santa Barbara
Dilberto Ferraren, Visayas State University
William Finzer, Concord Consortium
Brittany Fiore-Gartland, University of Washington
Julia Fisher, Coker College
Michael Franklin, University of Chicago
Roger French, Case Western Reserve University
Kimberly Gardner, Kennesaw State University
Sommer Gentry, U.S. Naval Academy
Tara Ghazi, University of California, Berkeley
Richard Gill, Brigham Young University
Shana Gillette, U.S. Agency for International Development
Juan Godoy, Universidad Nacional de Córdoba, Consejo Nacional de Investigaciones Cientificas y Técnicas
Greg Goins, North Carolina A&T State University
Robert Gould, University of California, Los Angeles
C.K. Gunsalus, University of Illinois, Urbana-Champaign
Mirsad Hadzikadic, University of North Carolina, Charlotte
Jim Hammerman, TERC

Michael Harris, Bunker Hill Community College
John Hathaway, Brigham Young University, Idaho
Adam Hughes, Pew Research
Kristin Hunter-Thomson, Rutgers, The State University of New Jersey
Ambra Hyskaj, National Association of Public Health Albania
Charles Isbell, Georgia Institute of Technology
Mark Jack, Florida A&M University
Vandana Janeja, National Science Foundation
Bob Jecklin, University of Wisconsin, La Crosse
Xia Jing, Ohio University
Jeremiah Johnson, University of New Hampshire
John Johnstone, University of Alabama, Birmingham
Ryan Jones, Middle Tennessee State University
Kari Jordan, Data Carpentry
Sungkyu Jung, University of Pittsburgh
Michael Kangas, Doane University
Nandini Kannan, National Science Foundation
Roxanne Kapikian, GlaxoSmithKline
Danny Kaplan, Macalester College
Casey Kennington, Boise State University
Sara Kiesler, National Science Foundation
Deepak Khatry, MedImmune
Brian Kotz, Montgomery College
Vladimir Krylov, Crimean Engineering and Pedagogical University
Kristin Kuter, Saint Mary's College
Jay Labov, National Academies of Sciences, Engineering, and Medicine
Paula Lackie, Carleton College
Sharon Lane-Getaz, St. Olaf College
Nicole Lazar, University of Georgia
Jeff Leek, Johns Hopkins University
Matthew Liberatore, University of Toledo
Haralambos Marmanis, Marmanis Group
Pat Marsteller, Emory University
Allison Master, University of Washington
Sorin Adam Matei, Purdue University
Abhinav Maurya, Carnegie Mellon University
Victoria McGovern, Burroughs Wellcome Fund
Daniel Angel Ferreira Mena, DAF-Engineering
Chris Mentzel, Gordon and Betty Moore Foundation
Sears Merritt, MassMutual Financial Group
Antoni Miklewski, Polish Academy of Sciences
Ashlea Milburn, University of Arkansas
Alex Montilla, U.S. Environmental Protection Agency

Sheri Morgan, Mental Health Association of Franklin and Fulton Counties
Richard Morris, MGI-RamCo
Mary Kehoe Moynihan, Cape Cod Community College
Bhramar Mukherjee, University of Michigan
Sherman Mumford, University of North Carolina, Charlotte
Ivo Neitzel, Faculdade de Ciências e Tecnologia de Birigui
Richard Nelesen, University of California, San Diego
Joseph Nelson, George Washington University
Claudia Neuhauser, University of Minnesota
Deborah Nolan, University of California, Berkeley
Kofi Nyamekye, Integrated Activity-Based Simulation Research, Inc.
Monika Oli, University of Florida
Fred Oswald, Rice University
Robert Panoff, Shodor
Dennis Pearl, Pennsylvania State University
Joan Peckham, University of Rhode Island
Vikas Pejaver, University of Washington
Gabriel Perez-Giz, National Science Foundation
Patrick Perry, New York University
Steve Pierson, American Statistical Association
Michael Posner, Villanova University
Earnestine Psalmonds-Easter, National Science Foundation
Hridesh Rajan, Iowa State University
Louise Raphael, Howard University
Renata Rawlings-Goss, Georgia Institute of Technology
Peggy Rejto, Normandale Community College
Loren Rhodes, Juniata College
Patrick Riley, Google, Inc.
Martina Rosenberg, University of New Mexico
Kim Roth, Juniata College
Bill Roweton, Chadron State College
Andee Rubin, TERC
Maya Sapiurka, Society for Neuroscience
Karl Schmitt, Valparaiso University
Kala Seal, Loyola Marymount University
Arun Sharma, Wagner College
Lauren Showalter, National Academies of Sciences, Engineering, and Medicine
Cláudio T. Silva, New York University
Christine Smith, University of New Mexico
Suzanne Smith, Johnson County Community College
S. Srinivasan, Texas Southern University

Anil Srivastava, Open Health Systems Laboratory
Natalya St. Clair, Concord Consortium
Victoria Stodden, University of Illinois, Urbana-Champaign
Martin Storksdieck, Oregon State University
George Strawn, National Academies of Sciences, Engineering, and Medicine
Ralph Stuart, Keene State College
Duncan Temple Lang, University of California, Davis
Kalum Udagepola, Scientific Research Development Institute of Technology Australia
Mel van Drunen, HAS University of Applied Sciences
William Yslas Velez, University of Arizona
Mladen Vouk, North Carolina State University
Ron Wasserstein, American Statistical Association
Cheryl Welsch, State University of New York, Sullivan
Mary Whelan, Arizona State University
Nekesha Williams, Louisiana State University
Emerald Wilson, Prince George's Community College
Brian Wingenroth, National Consortium for the Study of Terrorism and Responses to Terrorism, University of Maryland
William Winter, State University of New York College of Environmental Science and Forestry
Mary Wright, Brown University
Paul Zachos, Association for the Cooperative Advancement of Science and Education
Elena Zheleva, National Science Foundation

D

Data Science Oath

The committee proposed a Data Science Oath in its interim report and offers a revised version in this final report in Box D.2 (in parallel with a modern version of the Hippocratic Oath for physicians in Box D.1). Given the sensitive nature of certain types of data and the significant ethical implications of working with such data, similar efforts to establish a code of ethics for data scientists are under way throughout the field.[1] While these various codes at times intersect in their expressions of the ethical principles for data science, the committee hopes that its oath captures the gravity of data-driven decision making and provokes discussions on the future normative structure of data science.

While there are many common aspects to the Hippocratic Oath and the proposed Data Science Oath, there are also some key differences. Both oaths, however, share aspects of being necessary but not sufficient to address current and future ethical considerations.

The potential consequences of the ethical implications of data science cannot be overstated. Previously, data were small and specialized,

---

[1] To read about other work in the development of data science codes of ethics, see, for example, https://datapractices.org/community-principles-on-ethical-data-sharing/, http://datafordemocracy.org/projects/ethics.html, http://www.datascienceassn.org/code-of-conduct.html, http://www.rosebt.com/blog/open-for-comment-proposed-data-science-code-of-professional-conduct, https://dssg.uchicago.edu/2015/09/18/an-ethical-checklist-for-data-science/, http://thedataist.com/a-proposal-for-data-science-ethics/, https://www.accenture.com/t20160629T012639Z__w__/us-en/_acnmedia/PDF-24/Accenture-Universal-Principles-Data-Ethics.pdf, accessed January 31, 2018.

> **BOX D.1**
> **Hippocratic Oath**
>
> I swear to fulfill, to the best of my ability and judgment, this covenant:
>
> I will respect the hard-won scientific gains of those physicians in whose steps I walk, and gladly share such knowledge as is mine with those who are to follow.
>
> I will apply, for the benefit of the sick, all measures which are required, avoiding those twin traps of overtreatment and therapeutic nihilism.
>
> I will remember that there is art to medicine as well as science, and that warmth, sympathy, and understanding may outweigh the surgeon's knife or the chemist's drug.
>
> I will not be ashamed to say "I know not," nor will I fail to call in my colleagues when the skills of another are needed for a patient's recovery.
>
> I will respect the privacy of my patients, for their problems are not disclosed to me that the world may know. Most especially must I tread with care in matters of life and death. If it is given me to save a life, all thanks. But it may also be within my power to take a life; this awesome responsibility must be faced with great humbleness and awareness of my own frailty. Above all, I must not play at God.
>
> I will remember that I do not treat a fever chart, a cancerous growth, but a sick human being, whose illness may affect the person's family and economic stability. My responsibility includes these related problems, if I am to care adequately for the sick.
>
> I will prevent disease whenever I can, for prevention is preferable to cure.
>
> I will remember that I remain a member of society, with special obligations to all my fellow human beings, those sound of mind and body as well as the infirm.
>
> If I do not violate this oath, may I enjoy life and art, respected while I live and remembered with affection thereafter. May I always act so as to preserve the finest traditions of my calling and may I long experience the joy of healing those who seek my help.
>
> SOURCE: L.C. Lasagna, 1964, *Hippocratic Oath*, Modern Version, The Johns Hopkins Sheridan Libraries and University Museums. http://guides.library.jhu.edu/c.php?g=202502&p=1335759, accessed August 21, 2017.

but now data are pervasive. The fact that all humans are in this together (e.g., generating data and economic activity) means that they all have a responsibility to each other.

BOX D.2
Data Science Oath

I swear to fulfill, to the best of my ability and judgment, this covenant:

I will respect the hard-won scientific gains of those data scientists in whose steps I walk and gladly share such knowledge as is mine with those who follow.

I will apply, for the benefit of society, all measures which are required, avoiding misrepresentations of data and analysis results.

I will remember that there is art to data science as well as science and that consistency, candor, and compassion should outweigh the algorithm's precision or the interventionist's influence.

I will not be ashamed to say, "I know not," nor will I fail to call in my colleagues when the skills of another are needed for solving a problem.

I will respect the privacy of my data subjects, for their data are not disclosed to me that the world may know, so I will tread with care in matters of privacy and security. If it is given to me to do good with my analyses, all thanks. But it may also be within my power to do harm, and this responsibility must be faced with humbleness and awareness of my own limitations.

I will remember that my data are not just numbers without meaning or context, but represent real people and situations, and that my work may lead to unintended societal consequences, such as inequality, poverty, and disparities due to algorithmic bias. My responsibility must consider potential consequences of my extraction of meaning from data and ensure my analyses help make better decisions.

I will perform personalization where appropriate, but I will always look for a path to fair treatment and nondiscrimination.

I will remember that I remain a member of society, with special obligations to all my fellow human beings, those who need help and those who don't.

If I do not violate this oath, may I enjoy vitality and virtuosity, respected for my contributions and remembered for my leadership thereafter. May I always act to preserve the finest traditions of my calling and may I long experience the joy of helping those who can benefit from my work.